Optimal Process Design of 2-Phase and 3-Phase Production Separators

for Oil and Gas Installations

A Mathematical Formulation in Excel 2016

First published 2019

Revised and Updated 2020

Revised and Updated 2022

Copyright © 2019 - 2022 John T. Small.

All Rights Reserved

Disclaimer

The methods described in this book may be used by anyone desiring to do so, but the author shall not be held responsible or liable in any way for loss or damage resulting therefrom, or for the violation of any federal, state, or municipal regulations with which it may conflict.

Table of Contents

1. **INTRODUCTION** ... 6
2. **SEPARATOR CONFIGURATION AND DESIGN INPUTS** ... 8
   - 2.1 Separator Inlet Zone ... 8
     - 2.1.1 Dimensioning of Feed Pipe, Inlet Nozzle and Inlet Device ... 8
     - 2.1.2 Separator Flow Straightening Baffle ... 10
   - 2.2 Separator Gravity Settling Zone ... 10
     - 2.2.1 Size of Oil Droplets to be Removed from Gas (d_{OG}) ... 11
     - 2.2.2 Size of Gas Bubbles to be Removed from Oil (d_{GO}) ... 11
     - 2.2.3 Sizes of Water and Oil Droplets to be Removed from Liquid (d_{WO} & d_{OW}) ... 11
     - 2.2.4 Axial Length of Gravity Settling Zone ... 12
     - 2.2.5 Liquid Axial Velocity Considerations ... 13
   - 2.3 Dimensioning of Gas Outlet Nozzle and Gas Outlet Zone ... 14
     - 2.3.1 Gas Outlet Nozzle ... 14
     - 2.3.2 Gas Outlet Zone ... 14
   - 2.4 Dimensioning of Water Outlet Nozzle and Water Outlet Zone ... 14
     - 2.4.1 Water Outlet Nozzle ... 14
     - 2.4.2 Water Outlet Zone ... 15
   - 2.5 Dimensioning of Oil Outlet Nozzle and Oil Outlet Zone ... 15
     - 2.5.1 Oil Outlet Nozzle ... 15
     - 2.5.2 Oil Outlet Zone ... 16
   - 2.6 Oil and Water Retention Times ... 16
   - 2.7 Re-Entrainment of Oil Droplets into Gas Phase ... 17
   - 2.8 Slenderness Ratio ... 18
   - 2.9 Level Controls and Trips ... 18
     - 2.9.1 Level Control Intervals ... 18
     - 2.9.2 Unsteady State Operations – Slugging and Surging ... 19
     - 2.9.3 Level Trips ... 19
   - 2.10 Summary of Inputs to the Calculation Modules ... 21
   - 2.11 Navigating the Results of the Separator Design Calculations ... 24
3. **MATHEMATICAL FORMULATIONS AND OPTIMIZATION PROGRAMS** ... 25
   - 3.1 Calculation Flowcharts for Separator Optimization Programs ... 25
   - 3.2 Erosional Constraints on Nozzle Internal Diameters ... 28
   - 3.3 Determination of Feed Pipe / Inlet Nozzle Diameter ... 30
     - 3.3.1 Inlet Nozzle Diameter to Meet Velocity Head Criterion ... 30
     - 3.3.2 Impact of Inlet Nozzle Diameter on Vessel Sizing Calculations ... 31
   - 3.4 Determination of Gas Outlet Nozzle Diameter ... 31
     - 3.4.1 Gas Outlet Nozzle Diameter to Meet Velocity Head Criterion ... 32
   - 3.5 Determination of Liquid Outlet Nozzle Diameters ... 32
     - 3.5.1 Liquid Nozzle Diameters to Meet Velocity Criterion ... 32
     - 3.5.2 Impact of Liquid Outlet Nozzle Diameters on Vessel Sizing Calculations ... 33
   - 3.6 Sectional Area of Vessel and Volume of Heads Occupied by Liquid ... 33
   - 3.7 Gravity Settling Zone Calculations ... 33
     - 3.7.1 Droplet Settling Velocities ... 33

3.7.2	Length of Gravity Settling Zone L_{EFF}	34
3.7.3	Constraints on Oil and Water Retention Times	37
3.8	PREVENTION OF OIL RE-ENTRAINMENT INTO GAS PHASE	38
3.8.1	Model for Re-Entrainment of Oil Droplets into Gas Phase	39
3.8.2	Slenderness Ratio Constraint	40
3.9	AXIAL VELOCITY CONSIDERATIONS	41
3.10	FORMULATION OF LEVEL CONTROL INTERVALS	41
3.10.1	Interface Level Control Intervals	41
3.10.2	Oil Level Control Intervals	42
3.11	FOAMING ALLOWANCE	42
3.12	FORMULATION OF OIL TRIP SETTINGS	43
3.12.1	LLLL Trip Setting – 2-Phase Separator	43
3.12.2	HHLL Trip Setting – 2-Phase Separator	43
3.12.3	LLLL Trip Setting – 3-Phase Separator	43
3.12.4	HHLL Trip Setting – 3-Phase Separator	44
3.13	FREE ISSUE OF EXCEL WORKBOOK	44
4:	**CASE STUDY A: 2-PHASE PRODUCTION SEPARATOR DIMENSIONS FOR DROPLET CUT-OFF SIZES 140 MICRONS AND 300 MICRONS**	**45**
4.1	DEVELOPMENT OF SEPARATOR DESIGN PARAMETERS	45
4.2	KEY DESIGN CONSTRAINTS	46
4.3	CALCULATIONS AND RESULTS	46
4.4	SENSITIVITY CASE	47
5:	**CASE STUDY B: 3-PHASE PRODUCTION SEPARATOR DIMENSIONS FOR DIVERTER PLATE AND DIFFUSER INLET DEVICE OPTIONS**	**52**
5.1	METHODOLOGY	53
5.2	RESULTS	53
5.3	DISCUSSION OF RESULTS	55
5.4	CONCLUSIONS	56
6:	**CASE STUDY C: 3-PHASE PRODUCTION SEPARATOR DIMENSIONS ADEQUATE FOR PRODUCTION DURING EARLY AND LATE FIELD LIFE**	**62**
6.1	DEVELOPMENT OF SETTLING THEORY MODEL FOR EXISTING SEPARATOR	62
6.2	CALCULATE NOZZLE SIZES FOR NEW SEPARATOR	63
6.3	CALCULATE SEPARATOR VESSEL DIMENSIONS FOR LATE FIELD LIFE	64
6.4	CHECK SUITABILITY OF LATE LIFE VESSEL FOR EARLY LIFE FLUIDS	65
6.5	CONCLUSIONS	65
7:	**NOMENCLATURE**	**74**
8:	**REFERENCES**	**78**
9:	**ABOUT THE AUTHOR**	**79**

TABLE OF FIGURES

FIGURE 1: CONFIGURATION OF 2-PHASE SEPARATOR 5
FIGURE 2: CONFIGURATION OF 3-PHASE SEPARATOR 5
FIGURE 3: FLOWCHART FOR OPTIMIZED 2-PHASE SEPARATOR CALCULATION 26
FIGURE 4: FLOWCHART FOR OPTIMIZED 3-PHASE SEPARATOR CALCULATIONS 27
FIGURE 5: 2-PHASE SEPARATOR DESIGN PARAMETERS AND CONSTRAINTS 48
FIGURE 6: 2-PHASE SEPARATOR NOZZLE SIZING CALCULATIONS 49
FIGURE 7: 2-PHASE SEPARATOR DESIGN CALCULATION RESULTS – BASE CASE 50
FIGURE 8: 2-PHASE SEPARATOR DESIGN CALCULATION RESULTS – 300 MICRONS SENSITIVITY CASE 51
FIGURE 9 SHARED DESIGN PARAMETERS AND CONSTRAINTS - DIVERTER PLATE AND DIFFUSER OPTIONS 57
FIGURE 10 INLET NOZZLE CALCULATIONS - DIVERTER PLATE AND DIFFUSER OPTIONS 58
FIGURE 11 CALCULATION RESULTS - DIVERTER PLATE OPTION 59
FIGURE 12 CALCULATION RESULTS - DIFFUSER OPTION 60
FIGURE 13 THEORETICAL DROPLET CUT-OFF DIAMETERS - DIVERTER PLATE OPTION 61
FIGURE 14 THEORETICAL DROPLET CUT-OFF DIAMETERS – DIFFUSER OPTION 61
FIGURE 15: EXISTING 3-PHASE SEPARATOR DESIGN – L_{EFF} FOR OIL PHASE WITH 960 MICRONS WATER DROPLET CUTOFF SIZE 66
FIGURE 16: 3-PHASE SEPARATOR DESIGN – RESULTS OF CALCULATIONS FOR EXISTING SEPARATOR 67
FIGURE 17: 3-PHASE SEPARATOR DESIGN – NOZZLE SIZES BASED ON EARLY FIELD LIFE FLUID FLOWRATES 68
FIGURE 18: 3-PHASE SEPARATOR DESIGN – NOZZLE SIZES BASED ON LATE FIELD LIFE FLUID FLOWRATES 69
FIGURE 19: 3-PHASE SEPARATOR – INPUT DATA FOR LATE FIELD LIFE SIZING CALCULATION 70
FIGURE 20: 3-PHASE SEPARATOR – RESULTS OF DESIGN CALCULATIONS WITH LATE FIELD LIFE FLUIDS 71
FIGURE 21: 3-PHASE SEPARATOR – INPUT DATA FOR SIZING CHECK WITH EARLY FIELD LIFE FLUIDS 72
FIGURE 22: 3-PHASE SEPARATOR – RESULTS OF SIZING CHECK WITH EARLY FIELD LIFE FLUIDS 73

LIST OF TABLES

TABLE 1:-INPUTS FOR EROSION CALCULATIONS 23
TABLE 2: CONSTANTS AND VARIABLES FOR EROSION CALCULATIONS 29
TABLE 3 SALIENT RESULTS OF SEPARATOR CALCULATIONS 55
TABLE 4: FIELD DATA FOR 3-PHASE SEPARATOR STUDY 62
TABLE 5: CALCULATED NOZZLE SIZES FOR 3-PHASE SEPARATOR STUDY 64

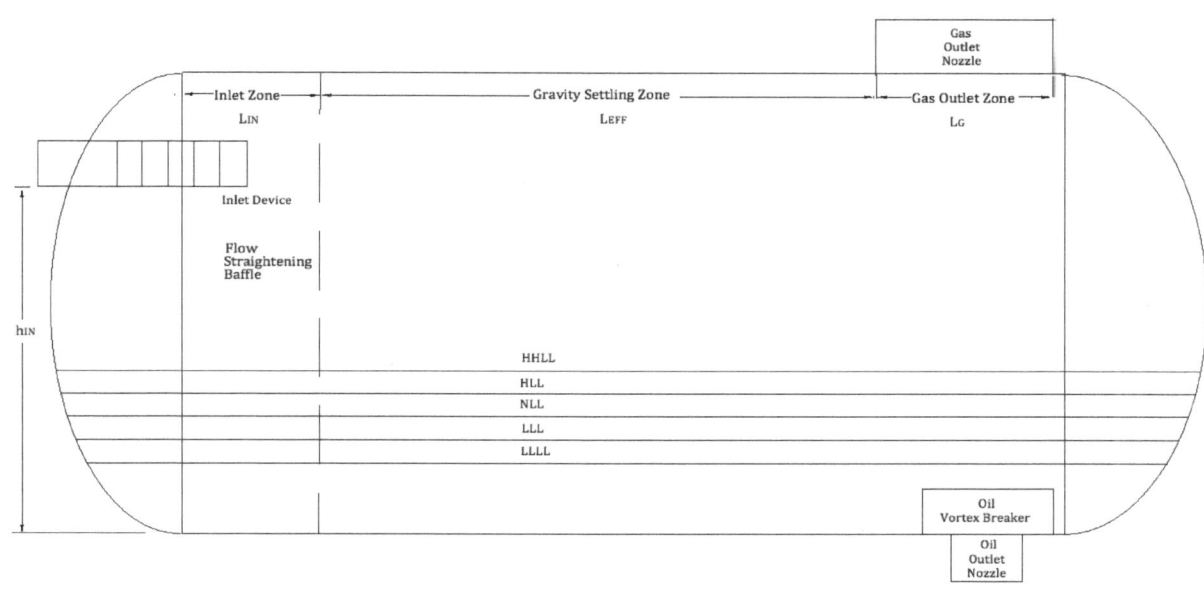

FIGURE 1: CONFIGURATION OF 2-PHASE SEPARATOR

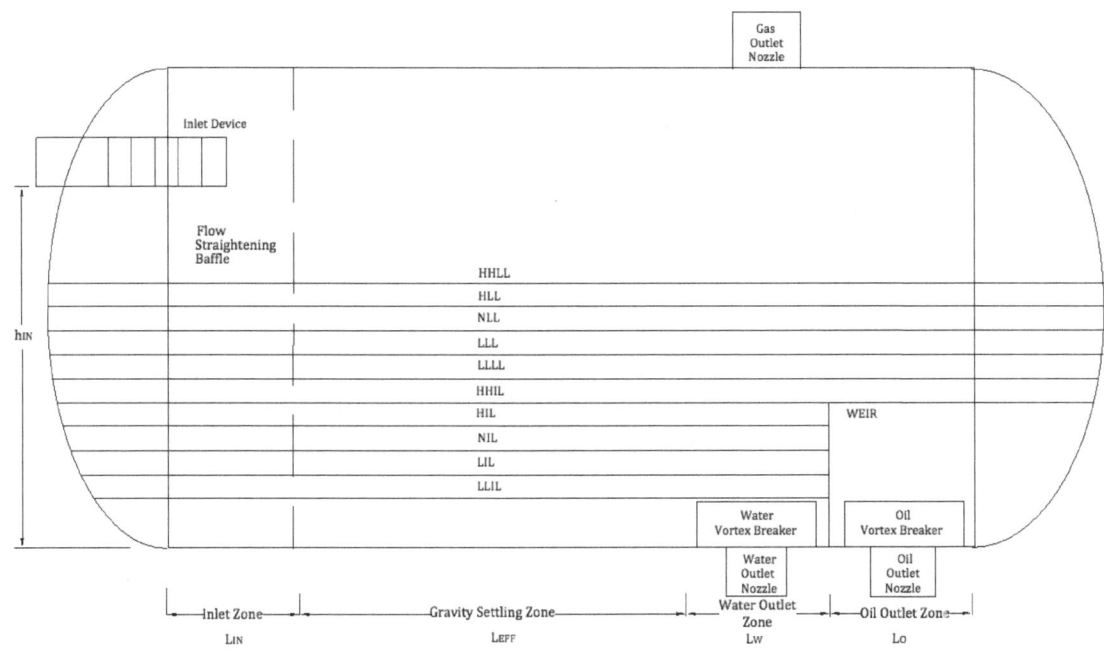

FIGURE 2: CONFIGURATION OF 3-PHASE SEPARATOR

1. INTRODUCTION

The 2- and 3-phase separator configurations which form the basis for this work are illustrated on Figure 1 and Figure 2 respectively.

The 3-phase separator is the flooded-weir type.

Both kinds of separator are commonly installed on oil and gas production installations.

They are relatively large, heavy vessels containing substantial hydrocarbon inventories. In the case of offshore installations, they utilize costly deck space and loading capacity. The associated pipework and valving are often bulky and challenging to accommodate in restricted module spaces. Modification projects especially may place hard limits on these factors.

At the same time, separator performance and control are often critical to successful operation of downstream equipment such as de-oiling, de-watering and gas compression units.

It follows that there is a strong incentive to minimize the size of the vessel and associated piping whilst achieving satisfactory performance in the separation, process control and safeguarding functions.

This optimization task requires resolution of key design aspects of the separator unit:

- Gravity settling zone dimensions for successful phase separation. Opposing schools of thought favor computation via droplet settling theory versus liquid retention times.
- Provision of adequate liquid hold up times and height differences between contiguous level control and trip settings.
- Type of inlet device, and its impact upon the size of inlet pipework and the dimensions of the separator vessel.
- Nozzle sizing according to erosional, velocity head and velocity criteria, and the impact on vessel dimensions.
- Maintaining axial velocities of oil and water phases within acceptable limits.
- Prevention of re-entrainment of oil droplets into the gas phase.
- Prevention of transient gas blowby or gross carryover of liquid during unplanned shutdown.
- Provision of adequate liquid and gas holdup volumes for slugging and surging.

- Prevention of spurious trips due to foaming, where applicable.

This book details a Microsoft® Excel spreadsheet which models numerically the factors listed above and optimizes the vessel dimensions using the Solver® add-in, based on the engineer's design input parameters and constraints. Macros are not employed, to maximize flexibility of use.

The principal calculation method for dimensioning the gravity separation zone of the vessel is based on settling theory, according to which droplets / bubbles of certain diameters settle to their respective continuous phases, however the spreadsheet allows the engineer to concurrently specify minimum allowable oil and / or water retention times if desired.

The objective function of the Solver algorithm may be set to minimize the volume of a separator vessel for the selected design parameters and constraints, or the engineer may choose other targets for optimization such as length of the gravity settling zone.

Case studies are presented which address the design of a 2-phase gas-condensate separator, and a 3-phase separator which is required to accommodate a major change in water cut during field life. For a 3-phase separator, the impact upon the key separator design aspects of installing a diverter plate versus a diffuser type inlet device is evaluated.

The case studies are founded on published field data.

Section 2 discusses development of the salient features of the separators, citing the technical references and criteria that have been considered. Key input design parameters and constraints are identified for solution of the dimensioning calculations.

Section 3 describes formulation of the spreadsheet calculation modules and how these are integrated to optimize the size of the separator vessel and nozzles.

Spreadsheets for the 2-phase and 3-phase separator design calculations are combined in a single workbook. Copies of the workbook are available from the author free issue to verified purchasers of this book.

2. SEPARATOR CONFIGURATION AND DESIGN INPUTS

Section 2 discusses the technical references and criteria that were considered in development of the salient features of the separators, and the input design parameters and constraints that are required for execution of the separator dimensioning calculations.

The topics are discussed under the following sequence of headings:

- Separator Inlet Zone
- Separator Gravity Settling Zone
- Gas Outlet Nozzle and Gas Outlet Zone
- Water Outlet Nozzle and Water Outlet Zone
- Oil Outlet Nozzle and Oil Outlet Zone
- Oil and Water Retention Times
- Re-entrainment of Oil Droplets into Gas Phase
- Slenderness Ratio
- Level Controls and Trips
- Summary of Required Inputs to the Calculation Modules
- Navigating the Results of the Separator Design Calculations

2.1 SEPARATOR INLET ZONE

The Inlet Zone components of both the 2- and 3-phase separators are generally defined in accordance with GPSA (2012) to include the feed pipe, inlet nozzle and inlet device. In addition, a flow straightening baffle is included downstream of the inlet device.

In compliance with GPSA and API 12J (2008), bulk phase separation is deemed to take place in the Inlet Zone during steady state operation.

2.1.1 DIMENSIONING OF FEED PIPE, INLET NOZZLE AND INLET DEVICE

The sizing calculations for these components require input to the Design Parameters in respect of flowrate, density and dynamic viscosity for each flowing phase, labelled Q_G; Q_O; Q_W; ϱ_G; ϱ_O; ϱ_W; μ_G; μ_O; μ_W.

GPSA (2012) recommends that the fluid velocity head J from a point 10 pipe diameters upstream of the inlet should match the requirements of the inlet device, to ensure good distribution of the phases and to minimize shattering of liquid droplets.

The feed pipe, inlet nozzle and inlet device are therefore considered to have the same internal diameter.

The engineer calculates component diameters which meet velocity head and erosional criteria, as described below. The spreadsheet selects from a look-up table the pipe diameter which satisfies the more stringent sizing criterion.

2.1.1.1 VELOCITY HEAD CONSIDERATIONS

The allowable velocity head J at the inlet device, and by extension in the feed pipe and inlet nozzle, depends on its design configuration.

Having selected a particular type of inlet device, the engineer inputs to the calculation panel for the inlet nozzle a value for allowable velocity head $J = \rho_M U_{N,IN}^{MAX^2}$.

GPSA (2012) cites typical allowable velocity head ranges for various types of inlet device:

J = 1500 – 3700 $[Pa]$ for half-pipe, v-baffle or elbow type.

J = 6000 – 9000 $[Pa]$ for diffuser type.

2.1.1.2 EROSION RATE CONSIDERATIONS

The size of the feed pipe, inlet nozzle and inlet device must also be great enough to satisfy erosional constraints during the service life of the separator.

In this work erosion rate is calculated in accordance with DNV GL (2015), who have developed models for prediction of erosive wear rates in the standard steel piping components found in oil and gas facilities. Grounded in the results of experimental investigations, the models enable sizing of pipework components based on the physical properties of the sand particles, the process fluid, and the pipework, and also take account of piping geometry.

It is assumed that from the perspective of erosion rate, the minimum diameter of the feed pipe / inlet nozzle / inlet device is governed by a pipe bend component in the feed pipe.

The engineer calculates the nozzle diameter required to meet the erosion target using Goal Seek function as per step number 6 in Section 2.10 below.

2.1.1.3 UPPER EDGE OF INLET NOZZLE RELATIVE TO TOP OF VESSEL

The upper edge of the inlet nozzle at h_{IN} should be located as close as permissible to the top of the vessel head, since this tends to minimize vessel diameter D_i. (Refer to Figure 1 and Figure 2). The proximity is however constrained by mechanical considerations to a certain proportion of D_i.

The proportion h_{IN} / D_i is defined by the engineer and is input to the Design Parameters as a value labelled R_{IN}^{IN}.

The value of R_{IN}^{IN} should ideally be obtained from a vessel specialist. Alternatively, a rule of thumb estimate of 0.8 may be used.

2.1.2 SEPARATOR FLOW STRAIGHTENING BAFFLE

The purpose of the flow straightening baffle downstream of the inlet device is to induce plug flow conditions in the Gravity Settling Zone, which is discussed further in Section 2.2.4 below. The location of the baffle relative to the vessel seam defines the axial length of the Inlet Zone and is labelled L_{IN}^{IN}.

The engineer inputs to the Design Parameters a fixed estimate for L_{IN}^{IN}.

2.2 SEPARATOR GRAVITY SETTLING ZONE

In this work, the principal calculation method for the phase separation aspect of separator sizing is based on settling theory, according to which droplets / bubbles of certain diameters settle to their respective continuous phases within the Gravity Settling Zone.

(Note: The calculation program also allows the engineer to concurrently specify minimum allowable oil and / or water retention times; refer to Section 2.6 and Case Study B below).

It is assumed that by designing the separator using settling theory for certain droplet sizes, all droplets of that size and larger settle out at terminal velocity to their respective continuous phase.

The gas phase holdup time in the Settling Zone needs to be sufficient for liquid droplets to settle out to the surface of the oil. Similarly, the holdup times of the oil and water phases need to be sufficient for settle out of gas bubbles and liquid droplets from the respective dispersed phase.

Noting that the terminal velocity for a given droplet size increases according to the density difference between the continuous and dispersed phases, Grødal and Realff (1999) highlight for each continuous phase the settling process which governs holdup time requirements in a 3-phase separator:

- In the continuous gas phase, settling out of oil droplets governs when oil density is closer to gas than is water density.
- In the continuous oil layer, settling of water droplets governs, since water density is closer to oil than is gas density.
- In the continuous water layer, flotation of the oil droplets governs, since oil density is closer to water than is gas density.

In the case of the 2-phase separator water is not present, so in the continuous oil phase only flotation of gas bubbles need be considered.

The engineer inputs to Design Parameters for each fluid phase the minimum diameter droplet that is to settle out in the Gravity Settling Zone:

d_{OG}° ; d_{GO}° ; d_{WO}° ; d_{OW}°.

The engineer calculates for each continuous phase the terminal settling velocity of dispersed droplets/bubbles using Goal Seek function, as per step number 4 in Section 2.10 below.

2.2.1 SIZE OF OIL DROPLETS TO BE REMOVED FROM GAS (D_{OG}°)

Arnold and Stewart (2008) cite field experience which suggests that selecting an oil droplet size $D_{OG}^{\circ} \equiv 140\ [\mu m]$ results in adequate oil removal from the gas phase.

2.2.2 SIZE OF GAS BUBBLES TO BE REMOVED FROM OIL (D_{GO}°)

GPSA (2012) states that selecting a gas bubble size $D_{GO}^{\circ} \equiv 200\ [\mu m]$ results in negligible gas carry-under to the oil outlet. (Note this is applicable for 2-phase separators only).

2.2.3 SIZES OF WATER AND OIL DROPLETS TO BE REMOVED FROM LIQUID (D_{WO}° & D_{OW}°)

The relative densities of the oil and water constituents being similar, the settling velocity of dispersed phase droplets is strongly influenced by the viscosity of the continuous phase. Oil viscosity is typically the order of 10 times that of water, therefore it is more often the allowable water droplet diameter that determines the dimension L_{EFF} for oil/water separation.

Arnold and Koszela (1990), writing from the perspective of design engineers, have reported good results sizing separators based on water droplet diameters D_{WO} = 500 – 1000 $[\mu m]$.

Arnold and Stewart (2008) provide estimated water-in-oil concentrations corresponding with this range of design water droplet diameter, for separator oil discharge streams.

This author analyzed production data from a paper by Laleh *et al* (2013) pertaining to an actual 3-phase separator operating satisfactorily on an offshore installation; oil from this separator is further processed sequentially in two downstream separators according to Hansen *et al* (1993).

Referring to Case Study C in the present work, the separator was found to be oil capacity limited, with a minimum water droplet diameter D_{WO} = 960 $[\mu m]$.

Arnold and Stewart (2008) suggest allowable oil droplet size D_{OW} = 200 $[\mu m]$ for flotation of oil from the water phase.

2.2.4 AXIAL LENGTH OF GRAVITY SETTLING ZONE

The axial length of the Gravity Settling Zone is labelled L_{EFF} $[m]$.

The balance of forces upon which settling theory is predicated applies strictly to horizontal plug flow of the continuous phase, as discussed in GPSA (2012), therefore it is necessary to define the serviceable length of the Gravity Settling Zone.

In the 2-phase separator, it is considered that the upwards flow component of gas towards the gas outlet nozzle would impede settling of oil droplets. Likewise, the downwards flow component of liquid to the oil outlet nozzle would impede rising gas bubbles. In practice the gas outlet nozzle will be much larger than the oil outlet nozzle. Thus, L_{EFF} is deemed to occupy the zone between the flow straightening baffle and the inboard edge of the Gas Outlet Zone.

Dimensioning of the Gas Outlet Zone is discussed in Section 2.3 below.

In the 3-phase separator, it is considered that the downwards flow component of liquid in the Water Outlet Zone would impede flotation of oil droplets and gas bubbles rising in the continuous liquid layers. Thus, L_{EFF} occupies the zone between the inlet flow straightening baffle and the inboard edge of the Water Outlet Zone.

Dimensioning of the Water Outlet Zone is discussed in Section 2.4 below.

The Gravity Settling Zone also contributes to other functions such as liquid holdup to meet Design Constraints on level control, surge / slug volumes, and where specified, liquid retention times. These functions may force extending L_{EFF} beyond what is required to accommodate the settling process.

The variable L_{EFF}^{TOTAL} $[m]$ is introduced to enable a solution for L_{EFF} which encompasses the multiple functions of the Gravity Settling Zone.

The engineer inputs to the calculation panel a trial value for L_{EFF}^{TOTAL}.

2.2.5 LIQUID AXIAL VELOCITY CONSIDERATIONS

The engineer inputs to Design Constraints maximum allowable values for water and oil axial velocity U_W^{MAX} and U_O^{MAX} $[m/s]$ within the Gravity Settling Zone.

These values affect the thickness of the respective layers and therefore influence L_{EFF}.

The engineer may wish to consider the following factors when setting values for U_O^{MAX} and U_W^{MAX}:

- GPSA (2012) mentions using a maximum axial velocity of 0.015 $[m/s]$ for the oil and water continuous phases, in a horizontal separator design with target droplet diameters of 150 $[\mu m]$.
- Analysis by this author of data pertaining to an actual offshore installation in a paper by Laleh *et al* (2013), indicates a 3-phase production separator operating satisfactorily with oil and water axial velocities of 0.2 and 0.1 $[m/s]$ respectively. The water droplet cutoff diameter was 960 $[\mu m]$. Refer to Case Study C in the present work.
- Arntzen (2016) writing from the PACO perspective advises consideration of liquid velocity upon the accuracy of level control instrumentation.

2.3 DIMENSIONING OF GAS OUTLET NOZZLE AND GAS OUTLET ZONE

2.3.1 GAS OUTLET NOZZLE

The gas outlet pipe and nozzle are assumed to have the same internal diameter. They are sized to meet the more stringent criterion of allowable velocity head and erosion rate. The engineer calculates component diameters which meet velocity head and erosional criteria, as described below. The spreadsheet selects from a look-up table the pipe diameter which satisfies the more stringent sizing criterion.

2.3.1.1 VELOCITY HEAD CRITERION

The engineer inputs to the calculation panel for the gas outlet nozzle, a value for allowable velocity head $J \equiv \rho_G U_{N,GAS}^{MAX\ 2}\ [Pa]$:

GPSA (2012) quotes industry guidelines of $J = 4500 - 5400\ [Pa]$ for gas outlet nozzles.

2.3.1.2 EROSION RATE CRITERION

From the perspective of erosion rate, it is assumed that the size of the gas outlet nozzle and outlet pipe is limited by a pipe bend component in the outlet pipe.

The engineer calculates the nozzle diameter required to meet the erosion target using Goal Seek function as per step number 6 in Section 2.10 below.

2.3.2 GAS OUTLET ZONE

For the 2-phase separator only, the Gas Outlet Zone L_G is deemed to extend vertically down through the phases and restricts the serviceable length of the Gravity Settling Zone L_{EFF}, as referred to above in Section 2.2.

The axial length of the Gas Outlet Zone is defined by the diameter of the gas outlet nozzle.

The outboard edge of the Gas Outlet Zone is assumed to be located on the vessel seam.

2.4 DIMENSIONING OF WATER OUTLET NOZZLE AND WATER OUTLET ZONE

2.4.1 WATER OUTLET NOZZLE

The water outlet pipe and nozzle are assumed to have the same internal diameter. They are sized to meet the more stringent criterion of allowable velocity and erosion rate. The engineer calculates nozzle diameters which meet velocity and erosional criteria, as described below. The spreadsheet selects from a look-up table the pipe diameter which satisfies the more stringent sizing criterion.

2.4.1.1 VELOCITY CRITERION

The engineer inputs to the calculation panel for the water outlet nozzle, a value for allowable velocity $U_{N,W}^{MAX}$.

The value may be selected from the ranges presented in API RP 14E (1991):

$0.9 \leq U^{MAX} \leq 4.6$ [m/s] for pipe between vessels.

$U^{MAX} = 0.9$ [m/s] for pipe into pump suction.

The lower limit on U^{MAX} is to prevent sand deposition.

2.4.1.2 EROSION RATE CRITERION

From the perspective of erosion rate, it is assumed that the size of the water outlet nozzle and outlet pipe is limited by a pipe bend component in the outlet pipe.

The engineer calculates the nozzle diameter required to meet the erosion target using Goal Seek function as per step number 6 in Section 2.10 below.

2.4.2 WATER OUTLET ZONE

Water Outlet Zone L_W is deemed to extend vertically up through the phases and restricts the serviceable length of the Gravity Settling Zone L_{EFF}, as referred to above in Section 2.2. Naturally this is applicable to the 3-phase separator only.

The axial length of the Water Outlet Zone is defined by the width of the vortex breaker on the water outlet, which the spreadsheet calculates to be twice the nozzle diameter as per Rochelle and Briscoe (2010).

The outboard edge of the Water Outlet Zone is assumed to abut the inboard side of the weir.

2.5 DIMENSIONING OF OIL OUTLET NOZZLE AND OIL OUTLET ZONE

2.5.1 OIL OUTLET NOZZLE

Dimensioning of the oil outlet nozzle is analogous to the water outlet nozzle.

The engineer inputs to the calculation panel for the oil outlet nozzle, a value for allowable velocity $U_{N,O}^{MAX}$.

The range of allowable velocity values is similar to the water outlet nozzle.

The engineer then calculates the oil outlet nozzle diameter required to meet the erosion target rate E^{MAX} using Goal Seek function as per step number 6 in Section 2.10 below.

2.5.2 OIL OUTLET ZONE

The Oil Outlet Zone is significant for the dimensions of the 3-phase separator only, extending axially from the outboard side of the weir to the vessel seam line.

The width of the Oil Outlet Zone is deemed to be the same as the oil outlet vortex breaker, which the spreadsheet calculates at twice the diameter of the oil outlet nozzle as per Rochelle and Briscoe (2010).

2.6 OIL AND WATER RETENTION TIMES

In this work, retention times for the liquid phases are evaluated in accordance with NORSOK P-002 (2014), using the respective normal levels downstream of the flow straightening baffle.

The outboard axial limits of the liquid volumes are demarcated by the vessel seam for the 2-phase separator and the weir plate for the 3-phase separator.

The engineer may specify values in Design Constraints for minimum required oil and water retention times $t_{RES,O}^{MIN}$ and / or $t_{RES,W}^{MIN}$. In this case the spreadsheet dimensions a separator which meets the more stringent criteria of retention times or droplet removal.

The engineer may have access to field data for separation based on oil and / or water retention times or they may wish to follow guidelines for retention times such as those referred to below.

For 2-phase separators, GPSA (2012) cites minimum oil retention time between 60 and 120 $[s]$ for adequate degassing, increasing with higher gas density or liquid viscosity.

Arnold and Stewart (2008) report generally successful separation with oil retention times between 30 and 180 [s], and a multiplication factor of 2 to 4 times to be applied when foaming is expected.

For 3-phase separators, sufficient oil retention time is required to allow coalescence of water droplets. Similarly, water retention time is required to allow coalescence of oil droplets.

Arnold and Stewart (2008) cite oil and water retention times ranging from 180 to 600+ [s] depending on the density and viscosity of the oil.

This author analyzed data reported by Laleh et al (2013) where a 3-phase separator operated successfully in the field with ca 60 [s] oil retention time. Oil density was 831.5 [kg/m^3] and viscosity was 5.25 E-3 [$Pa.s$] at operating temperature.

2.7 RE-ENTRAINMENT OF OIL DROPLETS INTO GAS PHASE

The following discussion applies to both 2-and 3-phase separators.

The cost of a vessel increases in line with the diameter, whereas excessive gas velocity results in re-entrainment of oil droplets into the gas phase. Thus, the scope for reduction of vessel diameter and cost may be constrained by the requirement to avoid re-entrainment.

This work uses the correlations which Ishii and Grolmes (1975) developed based on their experiments, to find the maximum velocity difference between the gas phase and the oil layer which precludes re-entrainment of oil droplets into the gas phase.

The spreadsheet design calculations constrain the diameter of the vessel to meet the maximum velocity difference based on liquid level at h_{HHLL}. This is a more conservative approach than setting the velocity constraint at h_{HLL}, since it minimizes the risk of liquid carryover during a level upset.

In practice the gas velocity constraint applies mainly to 2-phase separators. The diameter of 3-phase separators is more likely constrained by liquid velocity.

The engineer requires to input a value for oil-gas surface tension σ in Design Parameters.

2.8 SLENDERNESS RATIO

In traditional separator sizing methods as per Arnold and Stewart (2008), customarily a constraint is imposed on the maximum value of $\frac{L_{S-S}}{D_i}$ ("slenderness ratio") to avoid re-entrainment of oil into the gas phase.

The engineer may input a value for maximum allowable slenderness ratio R_S^{MAX} in Design Constraints when it is desired that the design calculations employ the slenderness ratio to ensure that the vessel diameter is sufficient to avoid re-entrainment of oil droplets into the gas.

The engineer may avoid activating the slenderness ratio constraint by inputting a suitably high value of R_S^{MAX} in Design Constraints.

Monnery and Svrcek (1994) provide guidelines for setting the maximum slenderness ratio as a function of operating pressure.

Viles (1993) provides a more detailed graphical approximation of maximum allowable slenderness ratio versus oil API gravity, for a range of separator operating pressures.

Viles indicates that these values of slenderness ratio are based on negligible oil velocity relative to gas, leading to a conservative evaluation of separator diameter.

2.9 LEVEL CONTROLS AND TRIPS

2.9.1 LEVEL CONTROL INTERVALS

Resolution of the level control and trip settings is a key part of the vessel sizing calculation.

Level control points are assigned at *LIL, NIL, HIL, LLL, NLL, HLL*; process trip points are assigned at *LLIL, HHIL, LLLL, HHLL*.

The height differences and holdup times between contiguous level control points and trips for the respective liquid layers are evaluated in compliance with NORSOK P-002 (2014).

For the purposes of level control calculations:

- The respective phases are considered to be single phase during steady state operations.

- When calculating liquid holdup times, credit is taken for volumes contained in the vessel heads.
- The more stringent requirement of height difference and holdup time governs the dimension of the control interval.

The engineer enters values in Design Constraints for the minimum acceptable height differences Δh_{CON}^{MIN} and holdup times Δt_{CON}^{MIN} between control levels.

NORSOK P-002 (2014) recommends values of 0.1 $[m]$ and 30 $[s]$ respectively.

The engineer inputs to the vessel calculation panel trial values for:

- h_{TV} ; h_{HHLL} ; h_{HLL} ; h_{NLL} ; h_{LLL} (2-phase separator) or
- h_{TV} ; h_{HHLL} ; h_{HLL} ; h_{NLL} ; h_{LLL} ; h_{LLLL} ; h_{HHIL}/h_{WEIR} ; h_{HIL} ; h_{NIL} ; h_{LIL} (3-phase separator)

2.9.2 UNSTEADY STATE OPERATIONS – SLUGGING AND SURGING

Transient intermittent flow conditions may occur in the feed pipe to the production separator, resulting in the arrival of alternating slugs of liquid and essentially dry surges of gas.

The ensuing movements in liquid level are deemed to be contained within the normal operating range between HLL and LLL, so that such conditions do not trigger level alarms.

The engineer may enter values in Design Constraints for the minimum acceptable volumes of slugs V_{SLUG}^{MIN} and surges V_{SURGE}^{MIN} to be accommodated within the control intervals between NLL – HLL and NLL - LLL respectively.

2.9.3 LEVEL TRIPS

It is assumed that level trip loops are designed with the appropriate SIL ratings and when initiated will cause closure of ESD valves in the separator inlet and outlet lines.

However, separator ESD valves are often large, requiring significant time to close. This may be considered when setting the trip levels.

2.9.3.1 LOW OIL LEVEL TRIP LLLL

The following discussion applies to both 2- and 3-phase separators.

Transient gas blow-by via the oil outlet line during closure of the ESD valve following *LLLL* trip, would potentially overpressure downstream equipment, or lead to damage of oil pumps due to low NPSH.

The engineer may enter a value in Design Constraints for the minimum acceptable holdup time $\Delta t_{SAF(LLLL)}^{MIN}$ between the *LLLL* trip and the top of the oil outlet vortex breaker sufficient to prevent gas blow-by during closure of the ESD valve.

Note that the spreadsheet calculates the top of the vortex breaker to be at a height equal to the diameter of the oil outlet nozzle, as per Rochelle and Briscoe (2010).

2.9.3.2 HIGH OIL LEVEL TRIP HHLL

The following discussion applies to both 2- and 3-phase separators.

Transient flooding of the separator inlet line during closure of the inlet ESD valve following *HHLL* trip, may cause gross carryover of oil into downstream gas processing equipment, leading to machinery damage.

The engineer may enter a value in Design Constraints for the minimum height difference $\Delta h_{SAF(HHLL)}^{MIN}$ between the *HHLL* trip and the lower edge of the inlet nozzle to meet the desired allowance for closure time of the inlet ESD valve.

2.9.3.3 INTERFACE LEVEL TRIPS LLIL AND HHIL

The following applies to 3-phase separator calculations only.

LLIL is set coincident with the top of the water outlet vortex breaker, since transient breakthrough of oil into the water treatment system while the inlet ESD valve is closing, would not lead to escalation of the incident.

Note that the spreadsheet calculates the top of the vortex breaker to be at a height equal to the diameter of the water outlet nozzle, as per Rochelle and Briscoe (2010).

Transient gas blow-by via the water outlet is not considered a credible scenario.

HHIL is set coincident with the top of the weir, since transient carryover of water into the oil compartment would not lead to escalation of the incident.

2.9.3.4 FOAMING ALLOWANCE

Foaming oil may potentially cause spurious operation of the *HHLL* trip.

The engineer may enter a value in Design Constraints for the minimum height allowance for foaming Δh_{FOAM}^{MIN} to impose a minimum allowable control interval between *HLL* and *HHLL*.

2.10 SUMMARY OF INPUTS TO THE CALCULATION MODULES

A user guide for the calculation spreadsheet is presented below. It is replicated within the Excel workbook.

1. Input values for Design Parameters.
2. Input values for Design Constraints.
3. Input trial values for level control and trip settings.
4. Calculate droplet terminal settling velocities for each phase, based on trial values of C_{Dd}^{TRIAL} and using Goal Seek function as follows:

   i. In the *Set Cell* box, enter the cell reference for the cell that contains the quotient $C_{Dd}^{TRIAL}/C_{Dd}^{Calc'd}$.
   ii. In the *To Value* box, enter the cell reference for the cell that contains the target value of 1 for the quotient.
   iii. In the *By Changing* Cell box, enter the cell reference for the cell that contains a trial value of C_{Dd}^{TRIAL}.

   Goal Seek updates the value of C_{Dd}^{TRIAL} to meet the target for equality with $C_{Dd}^{Calc'd}$, and provides the corresponding droplet terminal velocity values.

5. Input a trial value for length of gravity settling zone, L_{EFF}^{TOTAL}.
6. Calculate nozzle diameters required to meet erosion target, based on the trial value of D_E and the inputs listed on Table 1, using Goal Seek function as follows:

   i. In the *Set Cell* box, enter the cell reference for the cell that contains the equation for E^{MAX}.
   ii. In the *To Value* box, input the target erosion rate.
   iii. In the *By Changing Cell* box, enter the cell reference for the cell that contains a trial value of D_E.

Goal Seek updates the trial value of D_E^E to meet the engineer's erosional criterion, based on appropriate fluid flowrates and physical properties which are extracted automatically from the Design Parameters section.

7. Calculate the inlet and gas outlet nozzle diameters that are required to meet the engineer's velocity head targets.

8. Calculate the water and oil outlet nozzle diameters that are required to meet the engineer's velocity targets.

Using a look-up table, the spreadsheet selects nozzle diameters which meet the more stringent sizing criterion for each nozzle, and calculates vortex breaker sizes for the outlet nozzles.

9. Initiate calculation of the optimized separator dimensions using the Solver add-in as follows:

    i. In the *Set Objective* box, enter the cell reference for the cell that contains the function that is to be optimized (V, or L_{EFF}, etc).

    ii. In the *To* box, select *Max*, *Min* or *Value of*.

    NB: The *By Changing Variable Cells* box and the *Subject to Constraints* box should NOT be edited.

Variable	Description	Units
E^{MAX}	Maximum allowable erosion rate. Seek advice from Project Corrosion Specialist; alternatively, DNV GL (2015) suggests maximum total thickness loss of 0.5 [mm] during service life.	[mm/yr]
R	Radius of curvature of pipe, divided by nominal pipe diameter.	[-]
ρ_P	Density of pipe material.	[kg/m³]
GF	Geometry factor selected according to [4.3] in DNV GL (2015). Default value is 2.	[-]
ρ_S	Density of sand particles.	[kg/m³]
$ppmM_S$	Sand particle concentration in the fluid in parts per million on mass basis, during field life. NB: sand production often increases if the reservoir waters out with maturation. Seek advice from Reservoir Geologist or select a value from the range in DNV GL (2015).	[-]
d_S	Average sand particle diameter during field life. Seek advice from Reservoir Geologist or select a value from the range in DNV GL (2015).	[m]

TABLE 1: INPUTS FOR EROSION CALCULATIONS

[Back to top]

2.11 NAVIGATING THE RESULTS OF THE SEPARATOR DESIGN CALCULATIONS

The main panel of results for the design calculations shows the level control heights in semi-graphical format, with final values for the engineer's design constraints at the appropriate levels.

The upper part of the main panel shows final values for the dimensions of the vessel and nozzles.

Below the main panel are final values for the length of the Gravity Separation Zone, and the mass and volume of each phase of the vessel inventory.

3. MATHEMATICAL FORMULATIONS AND OPTIMIZATION PROGRAMS

Section 3 describes formulation of the various design calculation modules and discusses how these are integrated to optimize the size of the separators and nozzles.

3.1 CALCULATION FLOWCHARTS FOR SEPARATOR OPTIMIZATION PROGRAMS

Flowcharts of the mathematical programs for calculation of optimized 2-phase and 3-phase separator dimensions are shown below in Figure 3 and Figure 4 respectively.

The programs are developed in Excel 2016.

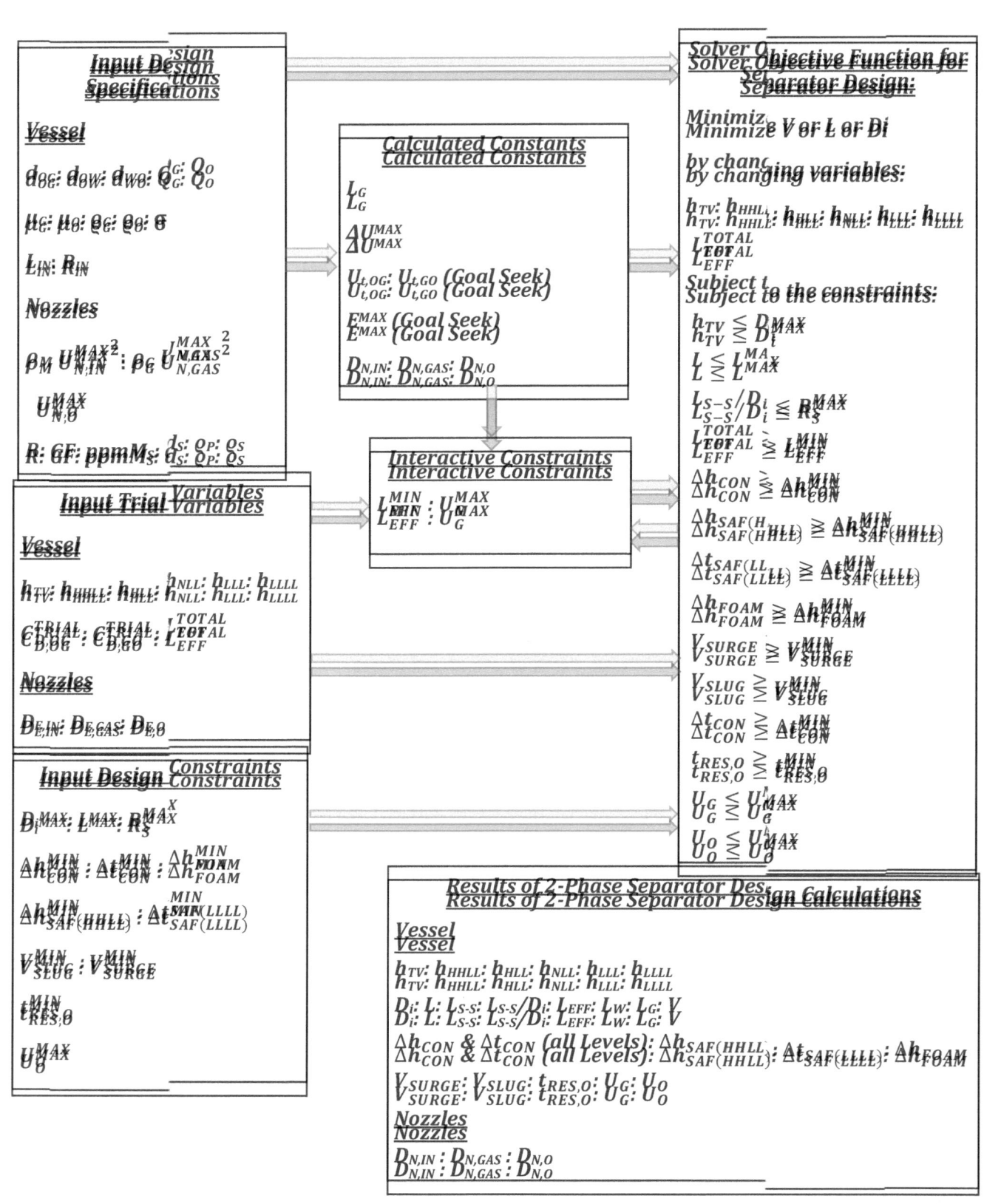

FIGURE 3: FLOWCHART FOR OPTIMIZED 2-PHASE SEPARATOR CALCULATION

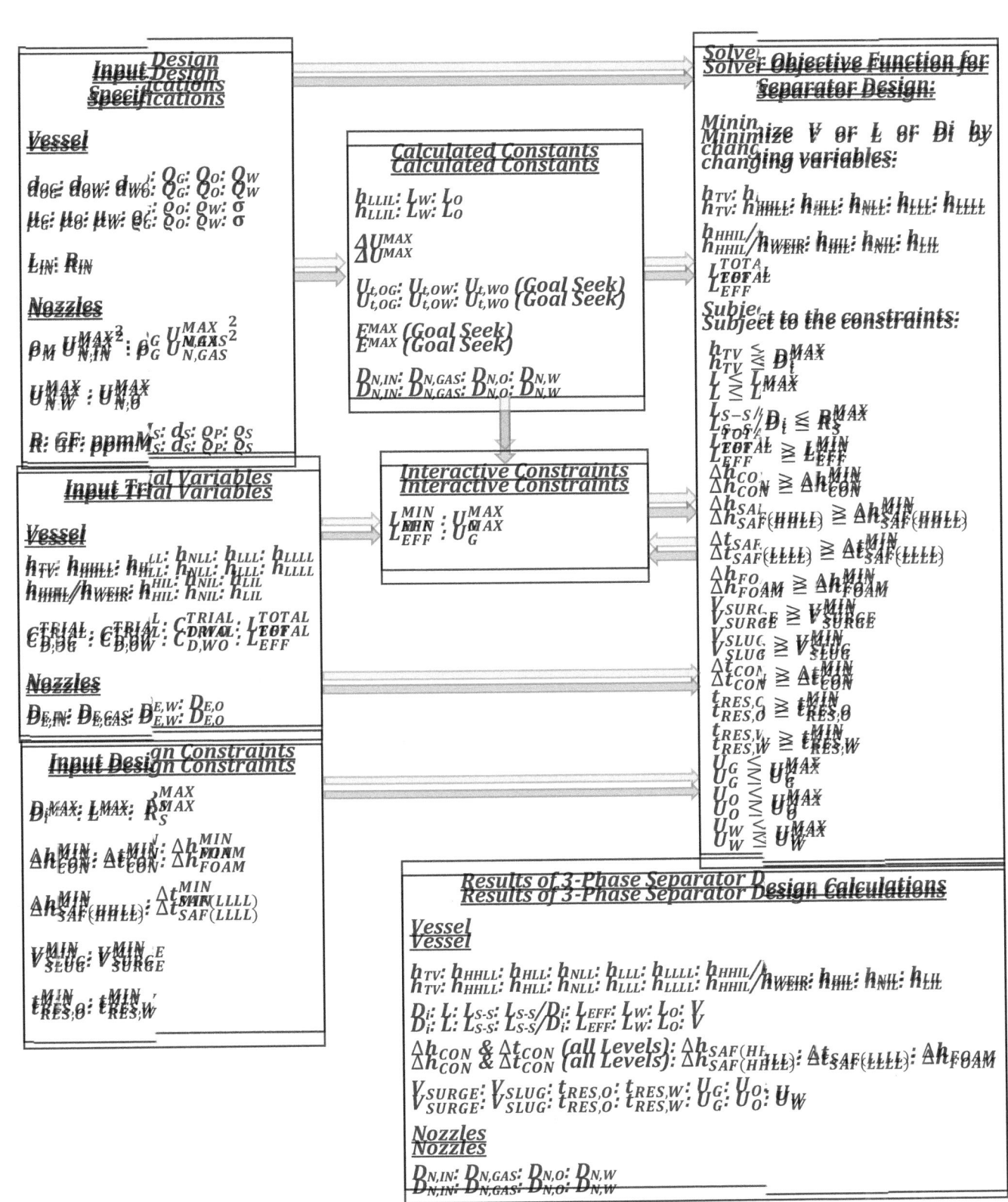

FIGURE 4: FLOWCHART FOR OPTIMIZED 3-PHASE SEPARATOR CALCULATIONS

3.2 EROSIONAL CONSTRAINTS ON NOZZLE INTERNAL DIAMETERS

The following discussion applies to all nozzles on both 2- and 3-phase separators.

According to DNV GL (2015), annual surface thickness loss due to erosion in a pipe bend can be written:

$$E^{MAX} = \frac{K.F(\alpha).U_S^n}{\rho_P.A_P}.G.C_1.GF.m_S.C_{unit} \quad [mm/year]$$

E^{MAX} is the maximum allowable annual surface thickness loss due to erosion.

To solve for E^{MAX} the spreadsheet employs the engineer's inputs in conjunction with the parameters listed in Table 2 below, which are embedded in the spreadsheet either as constants or calculated constants.

The engineer may input values for the highlighted parameters.

Iterative solution for E^{MAX} is required to calculate the pipe / nozzle diameter D_E that meets the target erosion rate.

This is accomplished in the spreadsheet using the Goal Seek feature as described above in Section 2.10, by modifying a trial value of D_E until convergence is achieved.

Symbol	Description	Unit
K, n	Material coefficients derived from testing for the combination of ductile steel and quartz sand.	[(m/s)-n]
F(α)	Correlates erosion rate with the impact angle of the sand on the pipe surface, which varies according to the radius of the pipe bend.	[–]
U_S	Impact velocity of the sand particles, equals fluid velocity.	[m/s]
ρ_P	Density of pipe material.	[kg/m³]
A_P	Pipe area exposed to erosion.	[m²]
G	Corrections function for sand particle diameter, according to the ratio of average sand particle diameter and pipe diameter; the angle of impact and velocity of sand particles; the density ratio of the fluid and the sand; fluid viscosity.	[–]
C_1	Model/geometry factor (=2.5). Accounts for multiple impact of the sand particles, concentration of particles at the outer part of the bend, and model uncertainty.	[–]
GF	Geometry factor accounting for deviation from model basis of straight upstream pipe length of minimum 10 pipe diameters.	[–]
m_S	Mass rate of sand particles.	[kg/s]
C_{unit}	Unit conversion factor = 3.15 E10.	[–]

TABLE 2: CONSTANTS AND VARIABLES FOR EROSION CALCULATIONS

3.3 DETERMINATION OF FEED PIPE / INLET NOZZLE DIAMETER

Having calculated the minimum required feed pipe / inlet nozzle diameter to meet the erosional criterion, which is labelled $D_{E,IN}$, the formulation shown in 3.3.1 below is used to calculate, based on the engineer's input value for $\rho_M U_{IN}^{MAX^2}$, the inlet nozzle diameter required to meet the velocity head criterion, which is labelled $D_{VH,IN}$.

The spreadsheet then, based on the larger of $D_{E,IN}$ and $D_{VH,IN}$, selects from a look-up table the nearest pipe size up. This is the inlet nozzle diameter, which is labelled $D_{N,IN}$.

3.3.1 INLET NOZZLE DIAMETER TO MEET VELOCITY HEAD CRITERION

The spreadsheet calculates as follows the minimum pipe / nozzle internal diameter which meets velocity head requirements, which is labelled $D_{VH,IN}$:

$$J \equiv \rho_M U_{IN}^{MAX^2} \qquad [Pa]$$

Giving:

$$U_{IN}^{MAX} \equiv \left(\frac{J}{\rho_M}\right)^{0.5} \qquad [m/s]$$

Then, for the 3-phase separator:

$$D_{VH,IN} \equiv \left(\frac{4(Q_G + Q_O + Q_W)}{\pi \left(\frac{J}{\rho_M}\right)^{0.5}}\right)^{0.5} \qquad [m]$$

Where:

$$\rho_M \equiv \frac{(Q_G \rho_G + Q_O \rho_O + Q_W \rho_W)}{(Q_G + Q_O + Q_W)} \qquad [kg/m^3]$$

And for the 2-phase separator:

$$D_{VH,IN} \equiv \left(\frac{4(Q_G + Q_O)}{\pi \left(\frac{J}{\rho_M}\right)^{0.5}}\right)^{0.5} \qquad [m]$$

Where:

$$\rho_M \equiv \frac{(Q_G \rho_G + Q_O \rho_O)}{(Q_G + Q_O)} \qquad [kg/m^3]$$

3.3.2 IMPACT OF INLET NOZZLE DIAMETER ON VESSEL SIZING CALCULATIONS

The following considerations apply to both the 2- and 3-phase separator calculations.

The upper edge of the inlet nozzle is located relative to the top of the vessel head h_{TV} according to the constraint R_{IN}.

Concurrently, the height of the LOWER edge of the inlet nozzle at h_{IN}, is constrained by the safety margin $\Delta h_{SAF(HHLL)}^{MIN}$ above the high liquid level trip at h_{HHLL} (see discussion in Section 3.12 below regarding $HHLL$ trip settings).

The value of h_{IN} remains in play during the vessel sizing calculation, to satisfy the following constraints:

$$h_{IN} \equiv R_{IN} \cdot D_i - D_{N,IN} \qquad [m]$$

And:

$$h_{IN} \geq h_{HHLL} + \Delta h_{SAF(HHLL)}^{MIN} \qquad [m]$$

3.4 DETERMINATION OF GAS OUTLET NOZZLE DIAMETER

The following considerations apply to both 2- and 3-phase separator calculations.

The minimum required gas outlet nozzle diameter to meet the erosional criterion is calculated in a similar way to the inlet nozzle, using gas properties only, and is labelled $D_{E,GAS}$.

The procedure described in 3.4.1 below is used to determine the gas outlet nozzle diameter required to meet the velocity head criterion, which is labelled $D_{VH,GAS}$.

The spreadsheet then, based on the larger of $D_{E,GAS}$ and $D_{VH,GAS}$, selects from a look-up table the nearest pipe size up. This is the gas outlet nozzle diameter, which is labelled $D_{N,GAS}$.

3.4.1 GAS OUTLET NOZZLE DIAMETER TO MEET VELOCITY HEAD CRITERION

The spreadsheet calculates as follows the minimum nozzle internal diameter which meets velocity head requirements, which is labelled $D_{VH,GAS}$:

$$J = \rho_G U_{GAS}^{MAX\,2} \qquad [Pa]$$

Giving:

$$U_{GAS}^{MAX} = \left(\frac{J}{\rho_G}\right)^{0.5} \qquad [m/s]$$

Then:

$$D_{VH,GAS} = \left(\frac{4Q_G}{\pi \left(\frac{J}{\rho_G}\right)^{0.5}}\right)^{0.5} \qquad [m]$$

3.5 DETERMINATION OF LIQUID OUTLET NOZZLE DIAMETERS

References to the water outlet nozzle apply only to 3-phase separator calculations.

The water and oil outlet pipe / nozzle diameters that are required to meet the erosional criterion are calculated in a similar way to the inlet nozzle, based on the appropriate single-phase properties, and are labelled $D_{E,W}$ and $D_{E,O}$

The procedure described in 3.5.1 below is used to determine the diameters that are required to meet the velocity criterion, which are labelled $D_{V,W}$ and $D_{V,O}$.

The spreadsheet, based on the larger of D_E and D_V for the respective outlet nozzles, selects from a look-up table the nearest pipe size up. These are the water and oil outlet nozzle diameters, which are labelled $D_{N,W}$ and $D_{N,O}$ respectively.

3.5.1 LIQUID NOZZLE DIAMETERS TO MEET VELOCITY CRITERION

Based on the selected values of U^{MAX} for the water and oil outlets, the spreadsheet calculates the minimum pipe / nozzle internal diameters to meet the velocity criterion, which are labelled $D_{V,W}$ and $D_{V,O}$ respectively.

3.5.2 IMPACT OF LIQUID OUTLET NOZZLE DIAMETERS ON VESSEL SIZING CALCULATIONS

The vortex breakers at the water and oil outlets are sized with an upstand equal to the nozzle internal diameter and a width of twice the nozzle diameter, as per Rochelle and Briscoe (2010).

The diameter of the water outlet nozzle of the 3-phase separator therefore:

- Locates the height of the *LLIL* trip at the top of the vortex breaker upstand as discussed in Section 2.9 above, establishing the foundation for the higher interface levels.
- Defines the axial dimension of the Water Outlet Zone, L_W.

The diameter of the oil outlet vortex breaker of the 3-phase separator defines the axial dimension of the Oil Outlet Zone, L_O.

3.6 SECTIONAL AREA OF VESSEL AND VOLUME OF HEADS OCCUPIED BY LIQUID

In this work A_L, the vertical cross-sectional area occupied by liquid, where the vessel internal diameter is D_i and the liquid height is h_L, is calculated according to the following equation, which is derived in Grødal and Realff (1999).

$$A_L = \left(h_L - \frac{D_i}{2}\right)(D_i h_L - h_L^2)^{0.5} + \frac{D_i^2}{4}\sin^{-1}\left(2\frac{h_L}{D_i} - 1\right) + \pi\frac{D_i^2}{8} \quad [m^2]$$

Semi-elliptical heads are assumed in this work, therefore the volume of the vessel head occupied by liquid at height h_L, can be written:

$$V_{H,h_L} = D_i^3 \frac{\pi}{24}\left\{3\left(\frac{h_L}{D_i}\right)^2 - 2\left(\frac{h_L}{D_i}\right)^3\right\} \quad [m^3]$$

3.7 GRAVITY SETTLING ZONE CALCULATIONS

3.7.1 DROPLET SETTLING VELOCITIES

When a spherical droplet or bubble is in vertical motion at its terminal velocity through a continuous medium, the drag force on it is in balance with the magnitude of its net weight:

$$F_d = \frac{\pi d_d^3}{6} |\rho_d - \rho_c| g \quad [N] \quad \ldots\ldots\ldots (1)$$

The correlation between the terminal velocity of droplets/bubbles with a certain diameter and density, and the physical properties of the continuous phase, is expressed in terms of the dimensionless groups for drag coefficient and Reynolds Number:

$$C_{Dd} \equiv \frac{8 F_d}{\pi d_d^2 U_t^2 \rho_c} \quad \ldots \ldots \ldots \ldots \ldots \ldots \ldots (2)$$

$$Re_d \equiv \frac{\rho_c U_t d_d}{\mu_c} \quad \ldots \ldots \ldots \ldots \ldots \ldots \ldots (3)$$

Arnold and Stewart (2008) provide a mathematical fit for the correlation:

$$C_{Dd} = \frac{24}{Re_d} + \frac{3}{Re_d^{0.5}} + 0.34 \quad \ldots \ldots \ldots (4)$$

Iterative solution is required to obtain the terminal velocity of the droplet/bubble, using the following steps:

1. Rearrange equation (2) to give:

$$U_t = \left(\frac{8 F_d}{\pi d_d^2 C_{Dd} \rho_c} \right)^{0.5} \quad [m/s] \ldots : (5)$$

2. Assume a trial value C_{Dd}^{Trial}.
3. Calculate U_t using C_{Dd}^{Trial} in equation (5).
4. Evaluate equations (3) and (4) with this value of U_t.

Repeat steps 2 – 4 with the updated value of C_{Dd}^{Trial} until it matches the value of $C_{Dd}^{Calc'd}$ calculated using equation (4). This corresponds with the correct value of U_t.

Iterative solution is accomplished in the spreadsheet using the Goal Seek feature with the test for equality $C_{Dd}^{Trial}/C_{Dd}^{Calc'd} \equiv 1$, as described in Section 2.10 above.

3.7.2 LENGTH OF GRAVITY SETTLING ZONE L_{EFF}

The spreadsheet calculates limiting values of L_{EFF} for settling droplets within each continuous phase, as described below in 3.7.2.1 - 3.7.2.4, then selects the governing value, i.e. that which is sufficient to allow the settling process in all of the continuous phases. This is labelled L_{EFF}^{MIN}.

Note that the value of L_{EFF}^{MIN} is a function of the thickness and cross-sectional areas of the phases, which vary during resolution of the Solver objective function.

The value of the variable L_{EFF}^{TOTAL}, which encompasses the multiple functions of the Gravity Settling Zone, is adjusted by Solver during resolution of the objective function, subject to the constraint $L_{EFF}^{TOTAL} \geq L_{EFF}^{MIN}$.

The value of L_{EFF}^{TOTAL} at solution of the objective function is reported as L_{EFF}.

3.7.2.1 LIMITS OF GAS CAPACITY ENVELOPE

The following discussion applies to both 2- and 3-phase separators.

The time taken t_G for the gas to traverse the distance $L_{EFF,G}$ must be no greater than the time required t_{OG} for oil droplets of a chosen diameter to settle from the top of the vessel to the oil surface, at terminal settling velocity $U_{t,OG}$.

It is considered that the effective operating level of the oil surface is between h_{LLL} and h_{HLL}.

The holdup time of the gas is maximized at h_{LLL}, as is the vertical distance through which the oil droplets must fall; the converse is true for the holdup time and distance parameters at h_{HLL}. These boundary conditions define the limits of the gas capacity envelope.

$$t_G \geq t_{OG} \qquad \left[\frac{s}{s}\right]$$

$$t_G \equiv \frac{Q_G}{V_G} \equiv \frac{Q_G}{L_{EFF,G}(A_{TV} - A_L)} \qquad \left[\frac{s}{s}\right]$$

And:

$$t_{OG} \equiv \frac{(h_{TV} - h_L)}{U_{t,OG}} \qquad \left[\frac{s}{s}\right]$$

Rearranging gives:

$$L_{EFF,G} \geq \frac{Q_G}{U_{t,OG}} \frac{(h_{TV} - h_L)}{(A_{TV} - A_L)} \qquad \left[\frac{m}{m}\right]$$

The limits of the gas capacity envelope at h_{HLL} and h_{LLL} can therefore be written:

$$L_{EFF,G} \geq \frac{Q_G}{U_{t,OG}} \frac{(h_{TV} - h_{HLL})}{(A_{TV} - A_{HLL})} \quad [m]$$

And:

$$L_{EFF,G} \geq \frac{Q_G}{U_{t,OG}} \frac{(h_{TV} - h_{LLL})}{(A_{TV} - A_{LLL})} \quad [m]$$

3.7.2.2 LIMITS OF OIL CAPACITY ENVELOPE – 2-PHASE SEPARATOR

The time taken by the oil to traverse the distance $L_{EFF,O}$ must be no greater than the time required for gas bubbles of a chosen diameter to rise from the bottom of the vessel to the oil-gas interface at terminal settling velocity $U_{t,GO}$, within the limits of the oil operating level between h_{HLL} and h_{LLL}.

The holdup time of the oil is maximized at h_{HLL}, as is the vertical distance through which the gas bubbles must rise; the converse is true for the holdup time and distance parameters at h_{LLL}. These boundary conditions define the limits of the oil capacity envelope.

The limits of the oil capacity envelope at h_{HLL} and h_{LLL} can therefore be written:

$$L_{EFF,O} \geq \frac{Q_O}{U_{t,GO}} \frac{h_{HLL}}{A_{HLL}} \quad [m]$$

And:

$$L_{EFF,O} \geq \frac{Q_O}{U_{t,GO}} \frac{h_{LLL}}{A_{LLL}} \quad [m]$$

3.7.2.3 LIMITS OF OIL CAPACITY ENVELOPE – 3-PHASE SEPARATOR

The time taken by the oil to traverse the distance $L_{EFF,O}$ must be no greater than the time required for water droplets of a chosen diameter to settle at terminal settling velocity $U_{t,WO}$ from the oil surface to the lowest operating oil-water interface, at h_{LIL}.

The operating limits of the oil level are considered to lie between h_{HLL} and h_{LLL}.

Oil holdup time and the vertical distance through which the water droplets must fall are both maximized at h_{HLL}; the converse is true for the holdup time and distance parameters at h_{LLL}. These boundary conditions define the limits of the oil capacity envelope.

The limits of the oil capacity envelope at h_{HLL} and h_{LLL} can therefore be written:

$$L_{EFF,O} \geq \frac{Q_o}{U_{t,WO}} \frac{(h_{HLL} - h_{LIL})}{(A_{HLL} - A_{LIL})} \quad [m]$$

And:

$$L_{EFF,O} \geq \frac{Q_o}{U_{t,WO}} \frac{(h_{LLL} - h_{LIL})}{(A_{LLL} - A_{LIL})} \quad [m]$$

3.7.2.4 LIMITS OF WATER CAPACITY ENVELOPE – 3-PHASE SEPARATOR ONLY

The time taken by the water to traverse the distance $L_{EFF,W}$ must be no greater than the time required for oil droplets of a chosen diameter to rise at terminal settling velocity $U_{t,OW}$ from the bottom of the vessel to the oil-water interface, within the limits of its operating level between h_{HIL} and h_{LIL}.

The holdup time of the water is maximized at h_{HIL}, as is the vertical distance through which the oil droplets must rise; the converse is true for the holdup time and distance parameters at h_{LIL}. These boundary conditions define the limits of the water capacity envelope.

The limits of the water capacity envelope at h_{HIL} and h_{LIL} can therefore be written:

$$L_{EFF,W} \geq \frac{Q_W}{U_{t,OW}} \frac{h_{HIL}}{A_{HIL}} \quad [m]$$

And:

$$L_{EFF,W} \geq \frac{Q_W}{U_{t,OW}} \frac{h_{LIL}}{A_{LIL}} \quad [m]$$

3.7.3 CONSTRAINTS ON OIL AND WATER RETENTION TIMES

Retention time calculations for the liquid phases are calculated in compliance with NORSOK P-002 (2014), using the respective normal levels downstream of the flow straightening baffle.

The outboard axial limits of the liquid volumes are demarcated by the vessel seam for the 2-phase separator and the weir plate for the 3-phase separator.

The constraint on oil retention times can be written as follows.

2-Phase Separator:

$$t_{RES,O} \equiv \frac{A_{NLL}(L_{EFF} \pm L_G)}{Q_O} \gtreqless t_{RES,O}^{MIN} \quad \left[\frac{s}{s}\right]$$

3-Phase Separator:

$$t_{RES,O} \equiv \frac{(A_{NLL} - A_{NIL})(L_{EFF} \pm L_W)}{Q_O} \gtreqless t_{RES,O}^{MIN} \quad \left[\frac{s}{s}\right]$$

The constraint on water retention time for the 3-Phase Separator can be written:

$$t_{RES,W} \equiv \frac{A_{NIL}(L_{EFF} \pm L_W)}{Q_W} \gtreqless t_{RES,W}^{MIN} \quad \left[\frac{s}{s}\right]$$

3.8 PREVENTION OF OIL RE-ENTRAINMENT INTO GAS PHASE

As described in 3.8.1 below, the spreadsheet calculates the maximum allowable axial velocity difference between the gas and oil phases ΔU^{MAX}, based on intensive properties of the two phases.

This work refers ΔU^{MAX} to liquid level at $HHLL$:

The gas velocity constraint for prevention of oil re-entrainment can be written:

$$U_G^{MAX} \leq \Delta U^{MAX} \pm U_{O,HHLL} \quad [m/s]$$

Where:

$$U_{O,HHLL} \equiv \frac{Q_O}{(A_{HHLL} - A_{NIL})} \quad [m/s]$$

The value of the gas velocity constraint U_G^{MAX} varies during resolution of the objective function, being dependent upon vessel diameter and the thickness of the continuous phases, which in turn may be influenced by the value of U_G^{MAX}.

The gas velocity limit for prevention of oil re-entrainment is satisfied when:

$$U_G \equiv \frac{Q_G}{(A_{TV} - A_{HHLL})} \leq U_G^{MAX} \quad [m/s]$$

3.8.1 MODEL FOR RE-ENTRAINMENT OF OIL DROPLETS INTO GAS PHASE

The re-entrainment model of Ishii and Grolmes (1975) is based on the Reynolds film number and interfacial viscosity number:

$$Re_f \equiv \frac{\rho_o U_o D_H}{\mu_o}$$

And:

$$N_\mu \equiv \frac{\mu_o}{\left\{\rho_o \sigma \left(\frac{\sigma}{g(\rho_o - \rho_g)}\right)^{0.5}\right\}^{0.5}}$$

Where wetted perimeter:

$$D_H \equiv \frac{4 A_{HHLL}^L}{P_{HHLL}^L} \qquad \left[\frac{m}{m}\right]$$

And hydraulic radius:

$$P_{HHLL} \equiv D_i \cos^{-1}\left(1 - 2\frac{h_{HHLL}}{D_i}\right) \qquad \left[\frac{m}{m}\right]$$

In accordance with Viles (1993), wetted perimeter and hydraulic radius are evaluated based on total liquid level, whereas U_o is evaluated based on the oil phase alone:

$$U_{o,HHLL} \equiv \frac{Q_o}{(A_{HHLL}^L - A_{NIL}^{NIL})} \qquad \left[m/s\right]$$

In terms of the Reynolds film number, Ishii and Grolmes identified three distinct regimes:

Low turbulence regime: $\{Re_f \leq 160\}$;

Transition regime: $\{160 \leq Re_f \leq 1635\}$;

Rough turbulent regime: $\{Re_f \geq 1635\}$

The tendency for re-entrainment increases with Reynolds film number.

For all Re_f regimes, the tendency for re-entrainment increases with higher values of N_μ, and exhibits a step change when N_μ exceeds 1/15 (0.067):

This work is based exclusively on the rough turbulent regime, since the Reynolds film number is expected to exceed the limit of 1635 by an order of 10. The calculated value of Re_f is however displayed in the results section for the purpose of validation.

Within the turbulent regime the full range of interfacial viscosity number is catered for.

The maximum allowable velocity difference between the gas phase and the oil layer which precludes oil re-entrainment, is formulated as follows for the rough turbulent regime, ($Re_f > 1635$):

$$\Delta U^{MAX} = \left(\frac{\sigma}{\mu_o}\right)\left(\frac{\rho_o}{\rho_g}\right)^{0.5} N_\mu^{0.8} \quad [m/s]$$

When $N_\mu \leq 0.067$.

And:

$$\Delta U^{MAX} = \left(\frac{\sigma}{\mu_o}\right)\left(\frac{\rho_o}{\rho_g}\right)^{0.5} \quad [m/s]$$

When $N_\mu > 0.067$.

3.8.2 SLENDERNESS RATIO CONSTRAINT

As discussed in 3.8.1 above, Ishii and Grolmes have shown empirically that in the rough turbulent regime, the allowable velocity difference between the gas phase and the oil layer:

- Is directly proportional to gas-oil surface tension
- Is an inverse function of gas density and dynamic viscosity

Surface tension decreases with pressure, whereas gas density and viscosity increase, and so with increasing separator pressure, larger vessel diameters are required to avoid re-entrainment of oil into the gas phase.

These considerations lead to the definition of maximum allowable slenderness ratio, $\frac{L_{S-S}}{D_i}$, which is denoted R_S^{MAX} in this work.

As discussed in Section 2.6 above, the engineer may impose a constraint on maximum allowable slenderness ratio in the spreadsheet calculation, to override the calculated constraint on gas velocity.

3.9 LIQUID AXIAL VELOCITY CONSIDERATIONS

The constraints which the engineer may impose upon maximum allowable oil and/or water velocity are calculated at *NLL* and *NIL* respectively, in accordance with GPSA (2012).

$$U_O = \frac{Q_O}{(A_{NLL} - A_{NIL})} \leq U_O^{MAX} \quad [m/s]$$

$$U_W = \frac{Q_W}{A_{NIL}} \leq U_W^{MAX} \quad [m/s]$$

3.10 FORMULATION OF LEVEL CONTROL INTERVALS

The engineer prescribes the minimum height differences Δh_{CON}^{MIN} and minimum holdup times Δt_{CON}^{MIN} required for level control intervals. These constraints are applied to all contiguous control levels in both the water and oil layers.

The spreadsheet calculates and selects for each control interval the more stringent criterion.

3.10.1 INTERFACE LEVEL CONTROL INTERVALS

Consideration of interface level control is required only for the 3-phase separator model.

Note that h_{LLIL} is a Calculated Constant, with a value equal to the diameter of the water outlet nozzle (see discussion in 3.5.2 above).

The minimum height difference for level control between h_{LIL} and h_{LLIL} can be written:

$$h_{LIL} - h_{LLIL} \geq \Delta h_{CON}^{MIN} \quad [m]$$

Since the axial limits for interface control intervals are the vessel inlet head and the weir, minimum holdup time for level control between h_{LIL} and h_{LLIL} can be written:

$$\frac{(V_{H,HIL} - V_{H,LLIL}) + (A_{LIL} - A_{LLIL})(L_{IN} + L_{EFF} + L_W)}{Q_W} \geq \Delta t_{CON}^{MIN} \quad [s]$$

Note that h_{HHIL} is located at weir height (see discussion in Section 2.9 above).

The minimum height differences and holdup times for level control between h_{NIL} - h_{LIL}; h_{HIL} - h_{NIL}; h_{HHIL} - h_{HIL} are therefore calculated using the pertinent values of h_L, A_L and $V_{H,L}$.

3.10.2 OIL LEVEL CONTROL INTERVALS

The minimum height difference for level control between h_{LLL} and h_{LLLL} can be written:

$$h_{LLL} - h_{LLLL} \geq \Delta h_{CON}^{MIN} \quad [m]$$

In the case of the 3-phase separator, the position of h_{LLLL} is deemed to be at weir height or above, therefore the control intervals for the oil layer extend across the entire length of the vessel, and the minimum holdup time for level control between h_{LLL} - h_{LLLL} can be written:

$$\frac{(2(V_{H,LLL} - V_{H,LLLLL}) + (A_{LLL} - A_{LLLLL})(L_{IN} + L_{EFF} + L_W + L_O))}{Q_O} \geq \Delta t_{CON}^{MIN} \quad [s]$$

For the 2-phase separator, there is no consideration required for water handling, and the minimum holdup time for level control between h_{LLL} - h_{LLLL} can be written:

$$\frac{(2(V_{H,LLL} - V_{H,LLLLL}) + (A_{LLL} - A_{LLLLL})(L_{IN} + L_{EFF} + L_G))}{Q_O} \geq \Delta t_{CON}^{MIN} \quad [s]$$

The minimum height differences and holdup times for level control between h_{NLL} - h_{LLL}; h_{HLL} - h_{NLL}; h_{HHLL} - h_{HLL} are calculated using the pertinent values of h_L, A_L and $V_{H,L}$.

3.11 FOAMING ALLOWANCE

The engineer may impose a foaming allowance between h_{HLL} and h_{HHLL} by setting a constraint $\Delta h_{FOAM}^{MIN} \geq 0$:

The spreadsheet applies the higher value of Δh_{FOAM}^{MIN} and Δh_{CON}^{MIN} for height difference in this control interval:

3.12 FORMULATION OF OIL TRIP SETTINGS

3.12.1 LLLL TRIP SETTING – 2-PHASE SEPARATOR

The engineer may set a constraint $\Delta t_{SAF(LLLL)}^{MIN} \geq 0$ for holdup time between LLLL trip and the top of the oil outlet vortex breaker:

The spreadsheet calculates oil hold-up time Δt_{SAF} above the vortex breaker according to:

$$\frac{\left(2(V_{H,LLLL} - V_{H,NO}) + (A_{LLLL} - A_{NO})(L_{IN} + L_{EFF} + L_G)\right)}{Q_O} \quad [s]$$

During resolution of the Solver objective function the LLLL trip setting is adjusted so that the oil holdup time meets the constraint $\Delta t_{SAF} \geq \Delta t_{SAF(LLLL)}^{MIN}$.

3.12.2 HHLL TRIP SETTING – 2-PHASE SEPARATOR

The engineer may set a constraint $\Delta h_{SAF(HHLL)}^{MIN} \geq 0$ on the height difference between h_{HHLL} and the lower edge of the inlet device at h_{IN}.

During resolution of the Solver objective function the HHLL trip setting is adjusted so that the height difference meets the constraint $\Delta h_{SAF} \geq \Delta h_{SAF(HHLL)}^{MIN}$.

The spreadsheet reports for information the corresponding oil holdup time Δt_{SAF}, which is calculated according to:

$$\frac{\left(2(V_{H,IN} - V_{H,HHLL}) + (A_{IN} - A_{HHLL})(L_{IN} + L_{EFF} + L_G)\right)}{Q_O} \quad [s]$$

3.12.3 LLLL TRIP SETTING – 3-PHASE SEPARATOR

The engineer may set a constraint $\Delta t_{SAF(LLLL)}^{MIN} \geq 0$ for holdup time between LLLL trip and the top of the oil outlet vortex breaker:

The spreadsheet calculates oil hold-up times Δt_{SAF} above the weir and in the oil compartment according to, respectively:

$$\frac{\left(2(V_{H,LLLL} - V_{H,HHIL}) + (A_{LLLL} - A_{HHIL})(L_{IN} + L_{EFF} + L_W + L_O)\right)}{Q_O} \quad [s]$$

And:

$$\frac{\left((V_{H,LLLL} - V_{H,NO}) + (A_{LLLL} - A_{NO})L_O\right)}{Q_O} \quad [s]$$

During resolution of the Solver objective function the *LLLL* trip setting is adjusted so that the sum of these oil holdup times meets the constraint $\Delta t_{SAF} \geq \Delta t_{SAF(LLLL)}^{MIN}$.

3.12.4 HHLL TRIP SETTING – 3-PHASE SEPARATOR

The engineer may set a constraint $\Delta h_{SAF(HHLL)} \geq 0$ on the height difference between h_{HHLL} and the lower edge of the inlet device at h_{IN}, to cater for the closure time of the inlet ESD valve.

During resolution of the Solver objective function the *HHLL* trip setting is adjusted so that the height difference meets the constraint $\Delta h_{SAF} \geq \Delta h_{SAF(HHLL)}^{MIN}$.

The spreadsheet reports for information the corresponding liquid holdup time Δt_{SAF}, which is calculated as shown below.

The combined liquid flowrate is used, since *HLL* is considered to be the upper limit of the operating envelope for liquid separation:

$$\frac{\left(2(V_{H,IN} - V_{H,HHLL}) + (A_{IN} - A_{HHLL})(L_{IN} + L_{EFF} + L_W + L_O)\right)}{(Q_O + Q_W)} \quad [s]$$

3.13 FREE ISSUE OF EXCEL WORKBOOK

The mathematical programs for calculation of optimized 2-phase and 3-phase separator dimensions are developed in Excel 2016 and are contained in a single workbook.

Copies of the Excel workbook are available free issue, by forwarding a receipt for purchase of this book to the author at the following address.

john@smallprocess.co.uk

4. CASE STUDY A: 2-PHASE PRODUCTION SEPARATOR DIMENSIONS FOR DROPLET CUT-OFF SIZES 140 MICRONS AND 300 MICRONS

It is desired to produce 10 MBD (Q_O = 0.0184 m³/s) condensate from an offshore gas-condensate reservoir.

In accordance with the recommendations cited in GPSA (2012):

- A horizontal production separator is selected since foaming is expected.
- A diffuser inlet device is selected to cope with the expected high gas rate with significant liquids.
  The diffuser device is known to minimize droplet fracture and promote gas distribution inside the separator.
  Diffuser devices have a relatively high maximum allowable mixed phase velocity head compared with other inlet devices, which allows minimization of the vessel inlet nozzle and inlet pipework.

4.1 DEVELOPMENT OF SEPARATOR DESIGN PARAMETERS

The hydrocarbon feed data below is adapted from Hoffmann *et al* (1953).

Laboratory tests have shown that based on a separator temperature of 201°F (94°C), peak condensate formation occurs at 1500 psig (103 barg); this operating pressure is therefore selected.

Coincident off-gas flowrate is 58904 SCF/BBL, based on gas density referred to standard conditions of 60°F and 14.7 psia, and condensate density ϱ_O = 675.3 kg/m³ measured at separator conditions.

Hydrocarbon analysis of the off-gas indicates molecular weight of 18.6 and flowing density ϱ_G = 66.61 kg/m³, giving a gas flowrate Q_G = 2.2787 m³/s at separator conditions.

For negligible carry-under of gas in the condensate outlet, the size of gas bubble to be separated d_{GO} = 200 μm, as recommended in GPSA (2012).

For the base case, the size of condensate droplet to be separated is set conservatively at d_{OG} = 140 μm.

A sensitivity case is carried out to investigate the impact of relaxing condensate separation to d_{OG} = 300-μm.

The objective function for the Solver design calculation is set to minimize vessel volume, to secure minimum hydrocarbon inventory and deck loading.

4.2 KEY DESIGN CONSTRAINTS

Maximum slenderness ratio is constrained to $R_S^{MAX} \equiv 4$, in compliance with the recommendations of Arnold and Stewart (2008).

The maximum allowable velocity head in the inlet nozzle/ inlet pipe $J \equiv \rho_M U_{N,IN}^{MAX^2} \equiv 7500$ kPa. This is a typical figure for a diffuser inlet device, according to GPSA (2012).

Margins between contiguous level control points and trips satisfy the more stringent condition of: height difference $\Delta h_{CON}^{MIN} \equiv 0.1$ m, and holdup time $\Delta t_{CON}^{MIN} \equiv 30$ s, in compliance with NORSOK P002 (2014).

A foam allowance $\Delta h_{FOAM}^{MIN} \equiv 0.25$ m is imposed on the control interval between HLL and HHLL.

A safety allowance $\Delta t_{SAF(LLLL)}^{MIN} \equiv 6$ s is imposed on the holdup time between LLLL and the top of the condensate outlet vortex breaker.

Separator nozzles are to be designed with maximum allowable erosion thickness loss $E^{MAX} \equiv 0.02$ mm/year.

4.3 CALCULATIONS AND RESULTS

Inputs and results of the production separator sizing calculations are reproduced below from the Excel spreadsheet output. The engineer's trial inputs are highlighted in yellow.

Separator Design Parameters and Constraints are shown on Figure 5.

Figure 6 shows the nozzle sizing calculations.

Erosion considerations are seen to govern the size of the inlet and gas outlet nozzles, which are both calculated at DN 641 mm (NPS 26 inch).

Figure 7 shows the main results of the production separator calculations.

The yellow highlighted cells show the final values of the trial inputs.

The vessel size is 2.53 m ID x 10.12 m S-S length, giving a slenderness ratio of 4.0.

Hydrocarbon inventory is 9441 kg.

It is noted that limiting the maximum slenderness ratio to a value of 4.0 results in U_G = 0.68 m/s, whereas according to the empirical method of Ishii and Grolmes, U_G^{MAX} = 1.29 m/s, the maximum allowable velocity to avoid re-entrainment of condensate into the gas phase. It might be judged therefore that there is scope for reducing the vessel diameter.

The value of L_{EFF} that is required for separation of condensate droplets from the gas phase is 8.48 m, compared with 0.68 m for separation of gas bubbles from the condensate phase. This indicates scope for reducing the length of the production separator in applications where larger condensate droplets may be allowed to exit with the gas phase, such as upstream of an inlet scrubber serving a compressor or dehydration unit.

4.4 SENSITIVITY CASE

A sensitivity case was carried out to investigate the impact of relaxing the condensate droplet size to d_{OG} = 300 μm from d_{OG} = 140 μm, with otherwise equal Design Parameters and Constraints.

The results of this calculation are shown in Figure 8. It is seen that the vessel ID reduces to 1.885 m, dictated by U_G^{MAX} = 1.297 m/s according to the method of Ishii and Grolmes.

The value of L_{EFF} that is required for separation of condensate droplets from the gas phase is reduced to 5.9 m, and vessel length is now 7.54 m S-S.

Hydrocarbon inventory is reduced by more than half, to 3622 kg.

Design Parameters

d_{GO}	2.00E-04	m	gas bubble size to be removed from oil
d_{OG}	1.40E-04	m	oil droplet size to be removed from gas
Q_G	2.2787	m³/s	gas flowrate at separator pressure and temperature
Q_O	0.0184	m³/s	oil flowrate at separator pressure and temperature
μ_G	1.70E-05	Pa.s	gas viscosity at separator pressure and temperature
μ_O	5.00E-04	Pa.s	oil viscosity at separator pressure and temperature
ϱ_G	66.61	kg/m³	gas density at separator pressure and temperature
ϱ_O	675.3	kg/m³	oil density at separator pressure and temperature
σ	0.02	N/m	gas/oil surface tension at separator pressure and temperature
L_{IN}	1.000	m	location of flow-straightening baffle, inboard from inlet tan line
R_{IN}	0.8	-	height of top of inlet nozzle expressed as a proportion of D_i

Design Constraints

D_i^{MAX}	4	m	maximum practicable internal diameter of vessel
L^{MAX}	20	m	maximum practicable overall length of vessel
R_S^{MAX}	4	-	maximum allowable slenderness ratio
Δh_{CON}^{MIN}	0.1	m	minimum height difference for level control between Normal Level / Level Alarm / Level Trip
Δt_{CON}^{MIN}	30	s	minimum holdup time between Normal Level / Level Alarm / Level Trip
Δh_{FOAM}^{MIN}	0.25	m	minimum height allowance for foaming, between HLL and HHLL
$\Delta h_{SAF(HHLL)}^{MIN}$	0.1	m	minimum height difference between HHLL and lower edge of inlet device, to avoid transient flooding of inlet device following HHLL trip
$\Delta t_{SAF(LLLL)}^{MIN}$	6	s	minimum holdup time between LLLL and top of oil vortex breaker, to avoid transient gas blowby following LLLL trip
V_{SLUG}^{MIN}	0	m³	minimum slug volume allowance between NLL and HLL
V_{SURGE}^{MIN}	0	m³	minimum surge volume allowance between NLL and LLL
$t_{RES,O}^{MIN}$	0	s	minimum retention time for oil at NLL, between flow-straightening baffle and vessel end

FIGURE 5: 2-PHASE SEPARATOR DESIGN PARAMETERS AND CONSTRAINTS

Back to 4.3

Calculation of Minimum Inlet Nozzle Size to Satisfy Erosional and Velocity Head Criteria

Erosional Rate (Pipe Bend)			Velocity Head		
E^{MAX}	0.020	mm/year	$\rho_M U^2{}_{NIN}{}^{MAX}$	7500	Pa
D_{EIN}	0.621	m	$U_{NIN}{}^{MAX}$	10.218	m/s
R	1.5	-	D_{VHIN}	0.535	m
ρ_P	7.80E+03	kg/m³		0.621	
GF	2	-			
ρ_S	2500	kg/m³	D_{NIN}	0.641	m
$ppmM_S$	30	-			
d_S	2.00E-04	m			

Calculation of Minimum Gas Outlet Nozzle Size to Satisfy Erosional and Velocity Head Criteria

Erosional Rate (Pipe Bend)			Velocity Head		
E^{MAX}	0.020	mm/year	$\rho_G U^2{}_{NGAS}{}^{MAX}$	5000	Pa
D_{EGAS}	0.619	m	$U_{NGAS}{}^{MAX}$	8.639	m/s
R	1.5	-	D_{VHGAS}	0.580	m
ρ_P	7.80E+03	kg/m³		0.619	
GF	2	-			
ρ_S	2500	kg/m³	D_{NGAS}	0.641	m
$ppmM_S$	30	-			
d_S	2.00E-04	m			

Calculation of Minimum Oil Outlet Nozzle Size to Satisfy Erosional and Velocity Criteria

Erosional Rate (Pipe Bend)			Velocity		
E^{MAX}	0.020	mm/year	$U_{NO}{}^{MAX}$	2.000	m/s
D_{EO}	0.073	m	D_{VO}	0.108	m
R	1.5	-		0.108	
ρ_P	7.80E+03	kg/m³			
GF	2	-			
ρ_S	2500	kg/m³	D_{NO}	0.154	m
$ppmM_S$	30	-			
d_S	2.00E-04	m			

FIGURE 6: 2-PHASE SEPARATOR NOZZLE SIZING CALCULATIONS

Back to 4.3

Results of Separator Design Calculations

D_i (m)	$D_{A,MM}$ (m)	$D_{A,MA}$ (m)	$D_{A,MAX}$ (m)	W_{SHAGE} (m³)	U_{LGA} (m/s)	Δt_{KAPI} (s)	L_L (m)	Δt_{KAPI} (s)	L_{L-SS} (m)	L_{S-S}/D_i	Δt_{FOMM} (m)	L_{LEFF} (m)	Δt_{CONV} (s)	Δh_{CONV} (m)	$L_{L,GAS}$ (m)	V_l (m³)
2.5380	0.0641	0.01534	0.0641				11.13855		10.01200	4.00000		8.34799			0.0641	55.51100
$t_{RES,DA}$ (s)				W_{SHAGE} (m³)		Δt_{KAPI} (s)		Δh_{KAPI} (m)			Δh_{FOMM} (m)		Δt_{CONV} (s)	Δh_{CONV} (m)		h_t (m)
								0.4448								2.5380
							678.8									1.3833
				0.68282												
										0.21050		35.353	0.21050			0.93935
			2.40008											0.10000		0.68585
45252												131.31				0.58585
				2.25757								128.23	0.10000			0.48585
						6.6						315.15	0.30322			0.16163
																0.15454

Trial Values	h_t (m)
H_{TV}	2.5380
h_{HHLL}	0.93935
h_{HLL}	0.68585
h_{NLL}	0.58585
h_{LLL}	0.48585
h_{LLLL}	0.16163

Calculation of Terminal Velocity of Gas Bubbles Rising in Oil

Goal Seek Calculation			Velocity Calculation		
$C_D/C_{D,GO}$ Trial	$C_D/C_{D,GO}$ calculated	$C_D/C_{D,GO}$ (Trial/Calculated)	$Re_{Re,OG}$	F_d/F_d (N/N)	$U_{t,bb}$ (m/s)
6.2R22	6.2121	1.0000	5.2626	2.50E-0808	0.001919

Calculation of Terminal Velocity of Oil Droplets Settling from Gas

Goal Seek Calculation			Velocity Calculation		
$C_D/C_{D,GO}$ Trial	$C_D/C_{D,GO}$ calculated	$C_D/C_{D,GO}$ (Trial/Calculated)	$Re_{Re,OG}$	F_d/F_d (N/N)	$U_{t,bb}$ (m/s)
1.0505	1.0505	1.0000	6962525	8.38E-0909	0.12626

Calculation of Maximum Allowable Actual Gas Velocity @ HHLL

U_G^{MAX} (m/s)	ΔU^{MAX} (m/s)	$N_A N_B$	Re_1	$P_{H,HHLL}$	$D_{H,HHLL}$
1.29090	1.27979	3.18E-0303	3.01E+0404	3.30606	2.04042

Calculation of d_{LEFF}/h_{TV} Values (m)

	@ h_{HLL}	@ h_{HHLL}
	0.6068	0.5059

Calculation of d_{LEFF}/h_{TV} Values (m)

	@ h_{HLL}	@ h_{HHLL}
	8.4848	8.4848

$L_{EFF,MIN}$	8.4848
L_{EFF} TOTAL	8.4848

Gas Inventory

(kg/kg)	(m³/h³)
303939	45.45030

Oil Inventory

(kg/kg)	(m³/h³)
646202	9.48080

Hydrocarbon Inventory

(kg/kg)	(m³/h³)
949441	55.55010

FIGURE 7: 2-PHASE SEPARATOR DESIGN CALCULATION RESULTS — BASE CASE

Back to 4.3

Results of Separator Design Calculations (300 microns sensitivity case)

D_i (m)	D_{MIN} (m)	D_{NOG} (m)	W_{SHRGE} (m³)	UL_{Gd} (m/s)	Δt_{KOH} (s)	LL (m)	L_{GAS} (m)	Δh_{KOH} (m)	L_{SSV/D_i}	L_{EFF} (m)	Δt_{CONV} (s)	Δh_{FOAM} (m)	L_{GAS5} (m)	Δh_{CONV} (m)	W (m³)
1.18855	0.6441	0.1584				8.04833	7.5340		4.4000	5.38999			0.6441		22.27977

t_{RESOV} (s)															h_i (m)
															1.18855
				1.29297		99.99		0.11818							0.86967
			1.31311							19.898		0.25050		0.25050	0.74949
										71.71				0.10000	0.49999
	1.5959														0.39999
			1.18787							65.65				0.10000	0.29999
										71.71				0.18232	0.16767
						6.6									0.15454

Trial Values

Trial Values	h_i (m)
h_{TV}	1.18855
h_{HHLL}	0.74949
h_{HLL}	0.49999
h_{NLL}	0.39999
h_{LLL}	0.29999
h_{LLLL}	0.16767

Calculated Terminal Velocity of Gas Bubbles Rising in Oil

Gas Side Calculation			Velocity Calculation		
$C_{D,60,OG}$ Trial	$C_{D,60,OG}$ calculated	$C_{D,60,OG}$ (Trial/Calcd)	$ReR_{G,OG}$	$F_{d,60}$ (N)N	$U_{t,60}$ (m/s)
6.2222	6.2121	1.0000	5.2626	2.50E-0808	0.01919

Calculated Terminal Velocity of Oil Droplets Settling from Gas

Gas Side Calculation			Velocity Calculation		
$C_{D,60,OG}$ Trial	$C_{D,60,OG}$ calculated	$C_{D,60,OG}$ (Trial/Calcd)	$ReR_{G,OG}$	$F_{d,60}$ (N)N	$U_{t,60}$ (m/s)
0.6060	0.6060	1.0000	28.73333	8.44E-0808	0.2444

Calculation of Minimum Allowable Axial Gas Velocity @ HHLL

$U_{G,G}^{MAX}$ (m/s)	AU_{G}^{MAX} (m/s)	$N_{v}N_{v}$	$ReRe_{v}$	P_{HHHLL}	D_{HHHLL}
1.1297	1.2795	3.31E-0303	3.86E+0404	2.5722	1.6083

Calculated h_{HLL} Values (m)

@ h_{HLL}	@ h_{HHLL}
0.9999	0.8080

Calculated h_{HHLL} Values (m)

@ h_{LLL}	@ h_{HHLL}
5.9090	5.8787

L_{HHmin}	5.9090
L_{EFF}^{TOTAL}	5.9090

Gas Inventory

(kg)	(m³)
128.888	19.154141

Oil Inventory

(kg)	(m³)
23.3334	3.45656

Hydrocarbon Inventory

(kg)	(m³)
362.822	22.275797

FIGURE 8: 2-PHASE SEPARATOR DESIGN CALCULATION RESULTS – 300 MICRONS SENSITIVITY CASE

Back to 4.3

5. CASE STUDY B: 3-PHASE PRODUCTION SEPARATOR DIMENSIONS FOR DIVERTER PLATE AND DIFFUSER INLET DEVICE OPTIONS

A 3-phase separator is to be designed for an offshore field development producing approximately 110,000 stock tank barrels of oil per day, plus associated off-gas and formation water.

The design work is based on the following component flow rates and key physical properties extracted from Grødal and Realff (1999):

Q_W	Q_O	Q_G	ϱ_W	ϱ_O	ϱ_G	μ_W	μ_O	μ_G	σ
m³/s	m³/s	m³/s	kg/m³	kg/m³	kg/m³	Pa.s	Pa.s	Pa.s	N/m
0.041	0.226	1.501	974.6	767.7	17.46	3.7E-4	7.3E-4	1.07E-5	0.02

Key design considerations are:

- The separator volume is to be minimized to secure minimum deck loading and hydrocarbon inventory.
- The size of the separator inlet piping is to be minimized to facilitate piping layout.

The impact upon the key separator design considerations of installing a diverter plate versus a diffuser type inlet device is to be evaluated.

5.1 METHODOLOGY

To enable valid comparison of the diverter plate versus the diffuser inlet device, discrete sizing calculations were carried for each Option, based on identical Design Parameters and Constraints, which are referenced below and listed in full on Figure 9.

The separators are sized to meet the following criteria cited for successful performance by Arnold and Stewart (2008):

- At least 3 minutes oil retention time.
- Separation of water droplets 500 µm and larger from the oil phase.
- Separation of oil droplets from the gas and water phases of 140 µm and 200 µm and larger, respectively.

The separators additionally satisfy the following specifications:

- Margins between contiguous level control points and trips satisfy the more stringent condition of: height difference Δh_{CON}^{MIN} = 0.1 m, and holdup time Δt_{CON}^{MIN} = 30 s, in compliance with NORSOK P002 (2014).
- Minimum holdup time between LLLL trip setting and top of oil outlet vortex breaker: $\Delta t_{SAF(LLLL)}^{MIN}$ = 10 s, to mitigate risk of transient gas blow-by.
- Maximum allowable erosion rate E^{MAX} = 0.02 mm/yr in the nozzles, evaluated in accordance with DNV GL (2015) based on a pipe bend component.
- Minimum height difference between HHLL trip and lower edge of inlet nozzle: $\Delta h_{SAF(HHLL)}^{MIN}$ = 0.1 m, to mitigate risk of transient flooding of nozzle.
- Height of upper edge of inlet nozzle expressed as proportion of vessel internal diameter, R_{IN} = 0.8.
- Slug and surge volumes V_{SLUG} and V_{SURGE} of at least 10 m³ each between NLL-HLL and NLL-LLL respectively.
- Maximum allowable axial velocity of oil: U_O^{MAX} = 0.2 m/s and water, U_W^{MAX} = 0.1 m/s. Values are calculated at NLL and NIL respectively, in compliance with GPSA (2012).

Maximum allowable velocity head values of J =2500 Pa and J = 7500 Pa were applied to sizing the deflector plate and diffuser inlet nozzle respectively, in accordance with typical values cited by GPSA (2012).

Each of the two designs engenders a particular final value of the settling zone length L_{EFF}^{TOTAL}, which in addition to accommodating settling of droplets, provides oil retention time.

The theoretical separation performance of the two designs was compared using settling theory to calculate oil droplet cut-off diameter in the gas and water phases and water droplet cut-off diameter in the oil phase.

These droplet cut-off diameters were obtained by iterative calculations to find the values of L_{EFF} matching L_{EFF}^{TOTAL}.

5.2 RESULTS

Salient results of the sizing calculations are presented on Table 3 below.

Figure 10 shows details of the nozzle sizing calculations for both Options.

Figure 11 and Figure 12 show details of the vessel design and hydrocarbon inventory calculation results for the Diverter Plate Option and Diffuser Option respectively.

Figure 13 and Figure 14 show the results of the final calculation iterations of theoretical oil and water droplet cut-off sizes for the Diverter Plate Option and Diffuser Option respectively.

		Diverter Plate Option	Diffuser Option
Inlet Nozzle Diameter, $D_{N,IN}$	m	0.743	0.591
Inlet Nozzle Diameter NB	ins	30	24
Vessel Diameter, D_i	m	3.82	3.57
Vessel Length, L_{S-S}	m	10.2	11.0
Vessel Volume, V	m³	132	122
Hydrocarbon Inventory	kg	44501	43084
Theoretical oil droplet cut-off diameter in gas phase, d_{OG}	μm	47	47
Theoretical oil droplet cut-off diameter in water phase, d_{OW}	μm	115	115
Theoretical water droplet cut-off diameter in oil phase, d_{WO}	μm	290	285

TABLE 3 SALIENT RESULTS OF SEPARATOR CALCULATIONS

Back to top

5.3 DISCUSSION OF RESULTS

In the case of the Diverter Plate Option, its relatively low allowable velocity head governs the 30" NB diameter of the inlet nozzle, whereas the higher allowable velocity head of the Diffuser Option nozzle results in a 24" NB inlet nozzle, matching the diameter imposed by the erosional constraint.

GPSA (2012) recommends separator inlet piping be the same diameter as the inlet nozzle for at least 10 pipe diameters, to avoid shattering of liquid droplets. The layout and weight implications of installing these large diameter pipes and associated valves and fittings favours the Diffuser Option due to its significantly smaller diameter.

The upper edge of the inlet nozzle is located at 0.8 x vessel diameter, whereas the lower edge of the inlet nozzle is specified to be 0.1 m above *HHLL*. These factors cause the Diverter Plate Option to require a vessel with larger diameter and volume than the Diffuser Option.

The larger volume of the Diverter Plate Option results in a hydrocarbon inventory 1400 kg larger than the Diffuser Option.

The 3 minutes oil retention time criterion governs the gravity settling zone length L_{EFF}^{TOTAL} for both Options, exceeding by a factor of 2 x the value of L_{EFF}^{MIN} that is required for settling water and oil droplets from the respective continuous liquid phase.

Theoretical droplet sizes that would be separated in the gravity settling zone are seen to be virtually identical for both Options according to settling theory, and significantly exceed the criteria for successful separation cited by Arnold and Stewart (2008).

5.4 CONCLUSIONS

The Diffuser Option requires a separator vessel with 10 m^3 less volume and containing 1400 kg less hydrocarbon than the Diverter Plate Option, based on providing 3 minutes oil retention time. Both Options achieve similar performance in terms of droplet cut-off diameters based on settling theory

The Diffuser Option requires 24" inlet pipework, compared with 30" in the case of the Diverter Plate Option, which is beneficial from the perspective of pipe routing and weight.

The Diffuser Option therefore offers a superior solution in terms of the key design considerations, without sacrificing separation performance.

Design Parameters

d_{OG}	1.40E-04	m	oil droplet size to be removed from gas
d_{OW}	2.00E-04	m	oil droplet size to be removed from water
d_{WO}	5.00E-04	m	water droplet size to be removed from oil
Q_G	1.5010	m^3/s	gas flowrate at separator pressure and temperature
Q_O	0.2260	m^3/s	oil flowrate at separator pressure and temperature
Q_W	0.0410	m^3/s	water flowrate at separator pressure and temperature
μ_G	1.07E-05	Pa.s	gas viscosity at separator pressure and temperature
μ_O	7.30E-04	Pa.s	oil viscosity at separator pressure and temperature
μ_W	3.70E-04	Pa.s	water viscosity at separator pressure and temperature
ϱ_G	17.46	kg/m^3	gas density at separator pressure and temperature
ϱ_O	767.7	kg/m^3	oil density at separator pressure and temperature
ϱ_W	974.6	kg/m^3	water density at separator pressure and temperature
σ	0.02	N/m	gas/oil surface tension at separator pressure and temperature
L_{IN}	1.000	m	location of flow-straightening baffle, inboard from inlet tan line
R_{IN}	0.8	-	height of top of inlet nozzle expressed as a proportion of D_I

Design Constraints

D_I^{MAX}	4	m	maximum practicable internal diameter of vessel
L^{MAX}	20	m	maximum practicable overall length of vessel
R_S^{MAX}	7	-	maximum allowable slenderness ratio
Δh_{CON}^{MIN}	0.1	m	minimum height difference for level control between Normal Level / Level Alarm / Level Trip, both liquid phases
Δt_{CON}^{MIN}	30	s	minimum holdup time between Normal Level / Level Alarm / Level Trip, both liquid phases
Δh_{FOAM}^{MIN}	0	m	minimum height allowance for foaming, between HLL and HHLL
$\Delta h_{SAFT(HHLL)}^{MIN}$	0.1	m	minimum height difference between HHLL and lower edge of inlet device, to avoid transient flooding of inlet device following HHLL trip
$\Delta t_{SAFT(LLLL)}^{MIN}$	10	s	minimum holdup time between LLLL and top of oil outlet vortex breaker, to avoid transient gas blowby following LLLL trip
V_{SLUG}^{MIN}	10	m^3	minimum slug volume allowance between NLL and HLL
V_{SURGE}^{MIN}	10	m^3	minimum surge volume allowance between NLL and LLL
$t_{RES,O}^{MIN}$	180	s	minimum retention time for oil, from flow-straightening baffle to weir, between NLL and NIL
$t_{RES,W}^{MIN}$	0	s	minimum retention time for water, from flow-straightening baffle to weir, between NIL and bottom of vessel
U_O^{MAX}	0.2	m/s	maximum allowable axial velocity of oil phase, based on cross-sectional area between NLL and NIL
U_W^{MAX}	0.1	m/s	maximum allowable axial velocity of water phase, based on cross-sectional area between NIL and bottom of vessel

FIGURE 9 SHARED DESIGN PARAMETERS AND CONSTRAINTS - DIVERTER PLATE AND DIFFUSER OPTIONS

Back to 5.1

Calculation of Minimum Inlet Nozzle Size to Satisfy Erosional and Velocity Head Criteria

Diverter Plate Inlet Device

Erosional Rate (Pipe-Bend)		
E^{MAX}	0.0020	mm/year
$D_{R,IN}$	0.05348	m
RR	1155	---
ρQ_P	7.70E+03	kg/m³
$GRGF$	2.2	---
ρQ_S	2.50E00	kg/m³
$ppm_{M,S}$	3030	---
$d_{d,S}$	2.00E-04	m

Velocity Head		
$\rho_M U^2_{N,IN}$	2500.0	Pa
$U^{MAX}_{N,IN}$	40.2394	m/s
$D_{N,IN}$	0.07224	m
	0.7241...	
$D_{N,IN}$	0.76243	m m

Diffuser Inlet Device

Erosional Rate (Pipe-Bend)		
E^{MAX}	0.0020	mm/year
$D_{R,IN}$	0.05348	m
RR	1155	---
ρQ_P	7.70E+03	kg/m³
$GRGF$	2.2	---
ρQ_S	2.50E00	kg/m³
$ppm_{M,S}$	3030	---
$d_{d,S}$	2.00E-04	m

Velocity Head		
$\rho_M U^2_{N,IN}$	75000	Pa
$U^{MAX}_{N,IN}$	77.4588	m/s
$D_{N,IN}$	0.0550	m
	0.5389...	
$D_{N,IN}$	0.55591	m m

FIGURE - ONE INLET NOZZLE CALCULATIONS DIVERTER PLATE AND DIFFUSER OPTIONS

[Back to 5.2](#)

FIGURE 10 CALCULATION RESULTS – DIVERTER PLATE OPTION

Results of Separator Design Calculations - Diffuser Inlet Device Installed

D_i (m)	$D_{N,IN}$ (m)	$D_{N,W}$ (m)	$D_{N,O}$ (m)	$D_{N,GAS}$ (m)	L (m)	$L_{S,S}$ (m)	$L_{S,S}/D_i$	L_{EFF} (m)	L_W (m)	L_O (m)	V (m³)
3.574	0.591	0.203	0.387	0.387	12.818	11.031	3.087	8.851	0.405	0.775	122.601

$t_{RES,O}$ (s)	$t_{RES,W}$ (s)	V_{SURGE} (m³)	V_{SLUG} (m³)	U_G (m/s)	U_O (m/s)	U_W (m/s)	Δt_{SAF} (s)	Δh_{SAF} (m)	Δh_{FOAM} (m)	Δt_{CON} (s)	Δh_{CON} (m)	h (m)	
												3.574	h_{TV}
								0.100				2.268	h_{IN}
				0.410			16			30	0.155	2.168	h_{HHLL}
					0.051				0.155			2.013	h_{HLL}
180			10.000							44	0.226	1.787	h_{NLL}
										44	0.226	1.562	h_{LLL}
		10.000								30	0.155	1.406	h_{LLLL}
						141						0.603	h_{HHLL}/h_{WEIR}
										68	0.100	0.503	h_{HIL}
										62	0.100	0.403	h_{NIL}
	140									55	0.100	0.303	h_{LIL}
						0.066				47	0.100	0.203	h_{LLIL}

Trial Values

	h (m)
h_{TV}	**3.574**
h_{HHLL}	**2.168**
h_{HLL}	**2.013**
h_{NLL}	**1.787**
h_{LLL}	**1.562**
h_{LLLL}	**1.406**
h_{HHLL}/h_{WEIR}	**0.603**
h_{HIL}	**0.503**
h_{NIL}	**0.403**
h_{LIL}	**0.303**

Calculated Terminal Velocity of Oil Droplets Settling from Gas

Goal Seek Calculation			Velocity Calculation		
$C_{D,OG}^{Trial}$	$C_{D,OG}^{calc'd}$	(Trial/Calc'd)	$Re_{d,OG}$	$F_{d,OG}$ (N)	$U_{t,OG}$ (m/s)
1.12	1.12	1.00	60.55	1.06E-08	0.265

Calculated $L_{EFF,G}$ Values (m)	
'@ h_{LLL}	1.96
'@ h_{HLL}	2.10

Calculated Terminal Velocity of Oil Droplets Rising in Water

Goal Seek Calculation			Velocity Calculation		
$C_{D,OW}^{Trial}$	$C_{D,OW}^{calc'd}$	(Trial/Calc'd)	$Re_{d,OW}$	$F_{d,OW}$ (N)	$U_{t,OW}$ (m/s)
6.72	6.72	1.00	4.79	8.50E-09	0.009

Calculated $L_{EFF,W}$ Values (m)	
'@ h_{LLL}	3.34
'@ h_{HLL}	2.64

Calculated Terminal Velocity of Water Droplets Settling from Oil

Goal Seek Calculation			Velocity Calculation		
$C_{D,WO}^{Trial}$	$C_{D,WO}^{calc'd}$	(Trial/Calc'd)	$Re_{d,WO}$	$F_{d,WO}$ (N)	$U_{t,WO}$ (m/s)
3.10	3.10	1.00	12.54	1.33E-07	0.024

Calculated $L_{EFF,O}$ Values (m)	
'@ h_{LLL}	3.14
'@ h_{HLL}	2.99

L_{EFF}^{MIN}	3.34
L_{EFF}^{TOTAL}	**8.85**

Calculation of Maximum Allowable Axial Gas Velocity @ HHLL

U_G^{MAX} (m/s)	ΔU^{MAX} (m/s)	N_{ut}	Re_T	P_{HHLL}	D_{LHHLL}
2.486	2.447	4.59E-03	1.65E+05	6.382	3.991

Gas Inventory		Oil Inventory		Total HC Inventory	
(kg)	(m³)	(kg)	(m³)	(kg)	(m³)
1070	61.285	42014	54.727	43084	116

FIGURE 12 CALCULATION RESULTS - DIFFUSER OPTION

Theoretical Droplet Cut-off Diameters in Separator - Diverter Plate Option

d_{OG}	4.70E-05	m	oil droplet size to be removed from gas
d_{OW}	1.15E-04	m	oil droplet size to be removed from water
d_{WO}	2.90E-04	m	water droplet size to be removed from oil

Calculated Terminal Velocity of Oil Droplets Settling from Gas

Goal Seek Calculation			Velocity Calculation			Calculated $L_{EFF,G}$ Values (m)	
$C_{D,OG}^{Trial}$	$C_{D,OG}^{calc'd}$	$C_{D,OG}^{(Trial/Calc'd)}$	$Re_{d,OG}$	$F_{d,OG}$ (N)	$U_{t,OG}$ (m/s)	'@ h_{LLL}	'@ h_{HLL}
6.68	6.68	1.00	4.82	4.00E-10	0.063	7.41	8.14

Calculated Terminal Velocity of Oil Droplets Rising in Water

Goal Seek Calculation			Velocity Calculation			Calculated $L_{EFF,W}$ Values (m)	
$C_{D,OW}^{Trial}$	$C_{D,OW}^{calc'd}$	$C_{D,OW}^{(Trial/Calc'd)}$	$Re_{d,OW}$	$F_{d,OW}$ (N)	$U_{t,OW}$ (m/s)	'@ h_{LIL}	'@ h_{HIL}
25.74	25.74	1.00	1.07	1.62E-09	0.004	8.32	6.51

Calculated Terminal Velocity of Water Droplets Settling from Oil

Goal Seek Calculation			Velocity Calculation			Calculated $L_{EFF,O}$ Values (m)	
$C_{D,WO}^{Trial}$	$C_{D,WO}^{calc'd}$	$C_{D,WO}^{(Trial/Calc'd)}$	$Re_{d,WO}$	$F_{d,WO}$ (N)	$U_{t,WO}$ (m/s)	'@ h_{LLL}	'@ h_{HLL}
9.69	9.70	1.00	3.13	2.59E-08	0.010	8.12	6.57

L_{EFF}^{MIN}	8.32
L_{EFF}^{TOTAL}	8.04

FIGURE 13 THEORETICAL DROPLET CUT-OFF DIAMETERS - DIVERTER PLATE OPTION

Back to 5.2

Theoretical Droplet Cut-off Diameters in Separator - Diffuser Option

d_{OG}	4.70E-05	m	oil droplet size to be removed from gas
d_{OW}	1.15E-04	m	oil droplet size to be removed from water
d_{WO}	2.85E-04	m	water droplet size to be removed from oil

Calculated Terminal Velocity of Oil Droplets Settling from Gas

Goal Seek Calculation			Velocity Calculation			Calculated $L_{EFF,G}$ Values (m)	
$C_{D,OG}^{Trial}$	$C_{D,OG}^{calc'd}$	$C_{D,OG}^{(Trial/Calc'd)}$	$Re_{d,OG}$	$F_{d,OG}$ (N)	$U_{t,OG}$ (m/s)	'@ h_{LLL}	'@ h_{HLL}
6.68	6.68	1.00	4.82	4.00E-10	0.063	7.93	8.85

Calculated Terminal Velocity of Oil Droplets Rising in Water

Goal Seek Calculation			Velocity Calculation			Calculated $L_{EFF,W}$ Values (m)	
$C_{D,OW}^{Trial}$	$C_{D,OW}^{calc'd}$	$C_{D,OW}^{(Trial/Calc'd)}$	$Re_{d,OW}$	$F_{d,OW}$ (N)	$U_{t,OW}$ (m/s)	'@ h_{LIL}	'@ h_{HIL}
25.74	25.74	1.00	1.07	1.62E-09	0.004	8.62	6.78

Calculated Terminal Velocity of Water Droplets Settling from Oil

Goal Seek Calculation			Velocity Calculation			Calculated $L_{EFF,O}$ Values (m)	
$C_{D,WO}^{Trial}$	$C_{D,WO}^{calc'd}$	$C_{D,WO}^{(Trial/Calc'd)}$	$Re_{d,WO}$	$F_{d,WO}$ (N)	$U_{t,WO}$ (m/s)	'@ h_{LLL}	'@ h_{HLL}
10.10	10.11	1.00	2.99	2.46E-08	0.010	8.74	7.16

L_{EFF}^{MIN}	8.85
L_{EFF}^{TOTAL}	8.85

FIGURE 14 THEORETICAL DROPLET CUT-OFF DIAMETERS – DIFFUSER OPTION

Back to 5.2

6. CASE STUDY C: 3-PHASE PRODUCTION SEPARATOR DIMENSIONS ADEQUATE FOR PRODUCTION DURING EARLY AND LATE FIELD LIFE

A 3-phase separator is to be designed for a new field development, taking account of the major increase in water flowrate which is predicted during field life. Production fluids data for early and late field life are shown on Table 4 below:

	Q_W	Q_O	Q_G	ρ_W	ρ_O	ρ_G	μ_W	μ_O	μ_G
	m³/s	m³/s	m³/s	kg/m³	kg/m³	kg/m³	Pa.s	Pa.s	Pa.s
Early	0.0797	0.5111	0.4556	1030	831.5	49.7	4.30E-4	5.25E-3	1.30E-5
Late	0.3456	0.3836	0.4556	1030	831.5	49.7	4.30E-4	5.25E-3	1.30E-5

TABLE 4: FIELD DATA FOR 3-PHASE SEPARATOR STUDY

An existing 3-phase separator with dimensions shown below is known to have processed satisfactorily production fluids having flowrates and physical properties like the early field life fluids.

It is therefore valid to design the new separator by extrapolation from a model of the existing separator.

The existing separator dimensions and production data are extracted from a paper by Laleh *et al* (2013) pertaining to an actual offshore installation:

Dimensions of existing separator: D_i = 3.328 m x L = 16.301 m.

   NLL: 1.664 m

   NIL: 0.625 m

6.1 DEVELOPMENT OF SETTLING THEORY MODEL FOR EXISTING SEPARATOR

To model the existing separator, it was necessary to calculate which continuous phase governs the separator capacity, and what is the droplet cut-off size of the dispersed phase.

The following design constraints were applied:

- Oil and water phase axial velocities U_O^{MAX} and U_W^{MAX} were constrained to 0.2 m/s and 0.1 m/s respectively, based on velocities calculated for early life liquid flows in the existing separator.
- Nozzle sizes were estimated based on appropriate velocity and velocity head constraints, and allowable erosion thickness loss $E^{MAX} \equiv 0.02$ mm/year.
- Level control settings were set in compliance with NORSOK P-002 (2014)

A preliminary run of the spreadsheet was made with oil and water droplet cutoff sizes of d_{OW} and d_{WO} of 500 µm. The terminal velocity values $U_{t,OW}$ and $U_{t,WO}$ were calculated using the Excel Goal Seek function operating on trial values of $\epsilon_{B,OW}$ and $\epsilon_{B,WO}$.

The results showed that the gravity settling zone for water droplets dispersed in the continuous oil phase was the order of 10 times longer than the other settling lengths. Oil capacity therefore governs the length of the existing separator.

Several iterations of the Solver spreadsheet were then made to find the water droplet cut-off size which resulted in a match for the separator overall length L. The Solver objective function was set to minimize the length of the gravity settling zone L_{EFF} and the vessel internal diameter D_i was constrained to a maximum value of 3.328 m, being the actual diameter of the existing vessel.

The results of the final iteration are shown on Figure 15 and Figure 16, where it is seen that 960 µm cutoff size for water droplets results in a value of 11.96 m for L_{EFF}, and the vessel length L is calculated at 16.318 m overall (cf 16.301 m actual).

Note that U_O at NLL reaches the maximum allowable value of 0.2 m/s, and U_G at HHLL is only 0.137 m/s compared with maximum allowable value of 1.10 m/s.

6.2 CALCULATE NOZZLE SIZES FOR NEW SEPARATOR

The nozzle internal diameters for the new separator must be adequate for the fluid flowrates in both early and late field life. Notably the inlet and water outlet nozzle diameters also have a significant impact on the dimensions of the separator vessel.

The results of nozzle sizing calculations for early and late field life fluid flowrates are shown on Figure 17 and Figure 18 respectively and summarized on Table 5 below.

		Early (mm)	Late (mm)	Selected (mm)
Inlet:	$D_{N,IN}$	591	692	692
Water Outlet:	$D_{N,W}$	255	489	489
Oil Outlet:	$D_{N,O}$	591	540	591
Gas Outlet:	$D_{N,GAS}$	305	305	305

TABLE 5: CALCULATED NOZZLE SIZES FOR 3-PHASE SEPARATOR STUDY

6.3 CALCULATE SEPARATOR VESSEL DIMENSIONS FOR LATE FIELD LIFE

The separator vessel dimensions required for late field life were calculated using the selected nozzle sizes in conjunction with the key design parameters and constraints shown on Figure 19.

Note that water flowrate has increased nearly 4-fold and oil flowrate has reduced by 25 %, compared with early field life.

Design constraint $\Delta t_{SAF(LLLL)}^{MIN}$ = 10 s holdup time between *LLLL* trip and top of oil vortex breaker has been incorporated to mitigate risk of transient gas blow-by.

The calculation results are shown on Figure 20 and were obtained by setting the Solver objective function to minimize *V* so as to secure minimum hydrocarbon inventory and deck loading.

Vessel dimensions are D_i = 4 m x L_{S-S} = 14.088 m.

The vessel diameter D_i is at the maximum allowable value of 4 m. This is forced by the water axial velocity U_W attaining its maximum allowable value of 0.1 m/s at *NIL*.

Although the relative flowrate of oil : water has drastically reduced, the oil capacity still governs gravity settling length L_{EFF}^{MIN} at 6.47 m versus 4.82 m required to meet water capacity. This is because the dynamic viscosity of the oil is the order of ten times that of the water, while the selected oil droplet size is more than half of the water droplet size.

The final length of the gravity settling zone L_{EFF} is however 11.03 m, driven by the minimum allowable holdup time between control levels Δt_{CON}^{MIN} = 30 s.

6.4 CHECK SUITABILITY OF LATE LIFE VESSEL FOR EARLY LIFE FLUIDS

As discussed above, the flowrate of water in late field life governs the vessel diameter D_i at 4 m and the length of the gravity settling zone L_{EFF} at 11.03 m.

The adequacy of these dimensions for processing early life fluids was checked using the appropriate design parameters and constraints shown on Figure 21, in conjunction with nozzle sizes in the "selected" column of Table 5.

The calculation results are shown on Figure 22 and were obtained by setting the Solver objective function to minimize L_{EFF}.

The calculation indicates that with a vessel diameter of 4 m, the length of the gravity settling zone is required to be 8.94 m to allow settling of water droplets from the oil phase. Since 11.03 m is available, the vessel sized for late field life is also suitable for early field life.

6.5 CONCLUSIONS

The foregoing calculations have shown that a separator of D_i = 4 m x L_{S-S} = 14.088 m is suitable for processing both early and late field life production fluids.

Separator dimensions are governed by oil capacity, considering droplet cut-off sizes of 500 μm (oil) and 960 μm (water).

The required nozzle sizes are (mm):

Inlet:	$D_{N,IN}$	692
Water Outlet:	$D_{N,W}$	489
Oil Outlet:	$D_{N,O}$	591
Gas Outlet:	$D_{N,GAS}$	305

Inlet nozzle size is based on diffuser type inlet device.

Design Parameters

$d_{o,max}$	1.40E+04	mm	oil droplet size to be removed from gas
$d_{o,max}$	5.00E+04	mm	oil droplet size to be removed from water
$d_{w,max}$	9.00E+04	mm	water droplet size to be removed from oil
QQ_G	0.8566	m³/s	gas flowrate at separator pressure and temperature
QQ_O	0.5111	m³/s	oil flowrate at separator pressure and temperature
QQ_W	0.0797	m³/s	water flowrate at separator pressure and temperature
μ_O	1.60E+05	Pa·s	gas viscosity at separator pressure and temperature
μ_O	5.25E+03	Pa·s	oil viscosity at separator pressure and temperature
μ_W	4.80E+04	Pa·s	water viscosity at separator pressure and temperature
ϱ_G	44.77	kg/m³	gas density at separator pressure and temperature
ϱ_O, ϱ_L	831.15	kg/m³	oil density at separator pressure and temperature
ϱ_W, ϱ_L	1030.30	kg/m³	water density at separator pressure and temperature
σ	0.020	N/m	gas/oil surface tension at separator pressure and temperature
L_{IN}	1.0000	m	location of flow distributor that flow fill the flow inlet is inlet line
$R_{IN}R_{IN}$	0.8 0.8	–	height of inlet nozzle expressed as a proportional D_t/D

Design Constraints

D_t MAX	3.32E+28	m	maximum practicable inner diameter of vessel
L MAX	20 20	m m	maximum practicable overall length of vessel
R MAX	7 7	–	maximum allowable slenderness ratio
Δh_{oil} MIN	0.1 0.1	m m	minimum length difference for liquid control between NLL / LLL / HLL / HHLL / LLLL / HHHL / etc.
Δt_{oil} MIN	30 30	s s	minimum hold-up time between flow level LLL / HLL / HHLL / LLLL / HHHL
$\Delta h_{w/o}$ MIN	0 0	m m	minimum length difference for flow level between LLL / HLL / HHLL / LLLL / HHHL
$\Delta h_{oil,w}$ MIN	0.1 0.1	m m	minimum length difference between HHLL / HHHL / etc. for oil outlet diversion to control flooding of inlet chute / etc. LLLL trip
$\Delta t_{oil,w}$ MIN	0 0	s s	minimum hold-up time between HHLL / HHHL / LLLL / etc. for oil outlet control between flow / etc. between NLL / HLL / HHLL trip
V_{surge} MIN	0 0	m³·m	minimum surge volume allowance between NLL / HLL / LLL
V_{surge} MIN	0 0	m³·m	minimum surge volume allowance between NLL / HLL / LLL
$t_{res,w,min}$ MIN	0 0	s s	minimum retention time for oil from flow inlet at the inlet flow inlet inlet in the water between NLL / HLL / LLL NIL
$t_{res,w,min}$ MIN	0 0	s s	minimum retention time for water from flow inlet at the inlet flow inlet inlet in the water between NLL / HLL / LLL NIL
U_O MAX	0.2 0.2	m/s	maximum allowable axial velocity of oil phase, based on cross-sectional area between NLL / HLL / LLL
U_W MAX	0.1 0.1	m/s	maximum allowable axial velocity of water phase, based on cross-sectional area between NLL / HLL / LLL

Calculated Terminal Velocity of Oil Droplet Settling from Gas

Goal Seek Calculation

$C_{D,G,O}$ Trial	$C_{D,G,O}$ calc.	(Trial-Calc.)²
0.990	0.990	1.0E0

Velocity Calculation

$Re_{G,O}$	$F_{d,G,O}$ (N)	$U_{G,O}$ (m/s)
959.97	1.10E-08	0.379

Calculated $L_{G,O,min}$ Values (m)

@ h/D	$L_{G,O,min}$ (m)
0.9 / h	0.991
	0.997

Calculated Terminal Velocity of Oil Droplet Rising in Water

Goal Seek Calculation

$C_{D,W,O}$ Trial	$C_{D,W,O}$ calc.	(Trial-Calc.)²
1.566	1.566	1.0E0

Velocity Calculation

$Re_{W,O}$	$F_{d,W,O}$ (N)	$U_{W,O}$ (m/s)
34.055	1.27E-07	0.0203

Calculated $L_{W,O,min}$ Values (m)

@ h/D	$L_{W,O,min}$ (m)
0.9 / h	1.83
	1.564

Calculated Terminal Velocity of Water Droplet Settling from Oil

Goal Seek Calculation

$C_{D,O,W}$ Trial	$C_{D,O,W}$ calc.	(Trial-Calc.)²	$Re_{O,W}$ calc.	$F_{d,O,W}$ (N)	$U_{O,W}$ (m/s)
12,402.47	12,402.48	1.0E0	2,362.36	9.03E-07	0.016

Calculated $L_{O,W,min}$ Values (m)

@ h/D	$L_{O,W,min}$ (m)
0.9 / h	11.96
	11.031.01

$L_{eff,min}$	11.96 .96
TOTAL	11.96 .96

FIGURE 5. EXISTING GAS-PHASE SEPARATOR DESIGN – 4 HP FOR OIL PHASE WITH 960 MICRONS WATER DROPLET CUTOFF SIZE

Back to 6.1

Results of Separator Design Calculations - Settling Theory Model for Vertical Separator Based on Actual Field Life Fluids

D1 (m)	DR,MIN (m)	DR,NOZ (m)	DR,ACAS (m)	LL (m)	L6,S (m)	L6,SS / D1	LL,HP (m)	LL,WH (m)	L0,H (m)	V (m³)
3.3288	0.5991	0.2555	0.3865	1.6388	1.4634	4.4033	1.1963	0.5599	1.1882	13.7123

tRESH (s)	VSHOUT (m³)	VSCUOUT (m³)	UL,d (m/s)	UL,wt (m/s)	UL,wt (m/s)	Δt₅ₛₘ (s)	Δh₅ₛₘ (m)	Δh₆ₐₘₙ (m)	Δt₆ₐₘₙ (s)	h (m)
										3.3288
										2.07071
			0.18737			9		0.29191	30	1.97971
		15.33333					0.10000			1.68181
62.62								0.29090	30	1.39391
	15.33333		0.20000			9		0.30000	30	1.09091
								0.30525	30	0.76666
					4					0.72929
				0.09090				0.10000	47	0.62929
1393.9								0.10000	44	0.52529
								0.10000	41	0.42929
								0.10474	61	0.25555

Calculation of Maximum Allowable Axial Gas Velocity @ HH-LLL

U₆ᴹᴬˣ (m/s)	U₆ᴰᴱˢ (m/s)	Re₆	N_μ	N_μ N_Re	P_m,HHLL	D_P,HHLL
1.10108	0.99393	6.6E+04	3.28E-02		5.83646	3.62672

Trial Values	hi (m)
hHW	3.3288
h_WHHUL	1.97971
h_HHL,HL	1.68181
h_NLH,NL	1.39391
h_LHL,LL	1.09091
h_LUHLL	0.76666
hvHUL/h_WIM	0.72929
h_HHL,HH	0.62929
h_NLh,NL	0.52529
h_LH,LL	0.42929

FIGURE 6.63 PHASE SEPARATOR DESIGN - RESULTS OF CALCULATIONS FOR EXISTING SEPARATOR

Back to 6.1

Calculation of Minimum Inlet Nozzle Size to Satisfy Erosional and Velocity Head Criteria

Erosional Rate (Pipe Bend)			Velocity Head		
E^{MAX}	0.020	mm/year	$\varrho_M U^2_{NIN}{}^{MAX}$	7500	Pa
$D_{E,IN}$	0.433	m	$U_{NIN}{}^{MAX}$	3.849	m/s
R	1.5	-	$D_{VE,IN}$	0.588	m
ϱ_P	7.80E+03	kg/m³		0.3881	
GF	2	-	$D_{N,IN}$	0.591	m

Calculation of Minimum Gas Outlet Nozzle Size to Satisfy Erosional and Velocity Head Criteria

Erosional Rate (Pipe Bend)			Velocity Head		
E^{MAX}	0.020	mm/year	$\varrho_G U^2_{NGAS}{}^{MAX}$	5000	Pa
$D_{E,GAS}$	0.301	m	$U_{NGAS}{}^{MAX}$	10.030	m/s
R	1.5	-	$D_{VE,GAS}$	0.240	m
ϱ_P	7.80E+03	kg/m³		0.3801	
GF	2	-	$D_{N,GAS}$	0.305	m

Calculation of Minimum Water Outlet Nozzle Size to Satisfy Erosional and Velocity Criteria

Erosional Rate (Pipe Bend)			Velocity		
E^{MAX}	0.020	mm/year	$U_{NW}{}^{MAX}$	2.000	m/s
$D_{E,W}$	0.143	m	$D_{V,W}$	0.225	m
R	1.5	-		0.2228	
ϱ_P	7.80E+03	kg/m³	$D_{N,W}$	0.255	m

Calculation of Minimum Oil Outlet Nozzle Size to Satisfy Erosional and Velocity Criteria

Erosional Rate (Pipe Bend)			Velocity		
E^{MAX}	0.020	mm/year	$U_{N,O}{}^{MAX}$	2.000	m/s
$D_{E,O}$	0.315	m	$D_{V,O}$	0.570	m
R	1.5	-		0.5708	
ϱ_P	7.80E+03	kg/m³	$D_{N,O}$	0.591	m

FIGURE 17: 3-PHASE SEPARATOR DESIGN – NOZZLE SIZES BASED ON EARLY FIELD LIFE FLUID FLOWRATES

Back to 6.2

Calculation of Minimum Inlet Nozzle Size to Satisfy Erosional and Velocity Head Criteria

Erosional Rate (Pipe Bend)

E_{MAX}	0.020	mm/year
$D_{E,IN}$	0.460	m
R	1.5	-
ϱ_P	7.80E+03	kg/m³
GF	2	-

Velocity Head

$\varrho_M U^2{}_{IN,MAX}$	7500	Pa
$U_{IN,MAX}$	3.569	m/s
$D_{VH,IN}$	0.650	m
	0.6438	
$D_{N,IN}$	0.692	m

Calculation of Minimum Water Outlet Nozzle Size to Satisfy Erosional and Velocity Criteria

Erosional Rate (Pipe Bend)

E_{MAX}	0.020	mm/year
$D_{E,W}$	0.274	m
R	1.5	-
ϱ_P	7.80E+03	kg/m³

Velocity

$U_{W,MAX}$	2.000	m/s
$D_{V,W}$	0.469	m
	0.4648	
$D_{N,W}$	0.489	m

Calculation of Minimum Gas Outlet Nozzle Size to Satisfy Erosional and Velocity Head Criteria

Erosional Rate (Pipe Bend)

E_{MAX}	0.020	mm/year
$D_{E,GAS}$	0.301	m
R	1.5	-
ϱ_P	7.80E+03	kg/m³
GF	2	-

Velocity Head

$\varrho_G U^2{}_{GAS,MAX}$	5000	Pa
$U_{GAS,MAX}$	10.030	m/s
$D_{VH,GAS}$	0.240	m
	0.2901	
$D_{N,GAS}$	0.305	m

Calculation of Minimum Oil Outlet Nozzle Size to Satisfy Erosional and Velocity Criteria

Erosional Rate (Pipe Bend)

E_{MAX}	0.020	mm/year
$D_{E,O}$	0.277	m
R	1.5	-
ϱ_P	7.80E+03	kg/m³

Velocity

$U_{O,MAX}$	2.000	m/s
$D_{V,O}$	0.494	m
	0.4824	
$D_{N,O}$	0.540	m

FIGURE 18: 3-PHASE SEPARATOR DESIGN – NOZZLE SIZES BASED ON LATE FIELD LIFE FLUID FLOWRATES

Design Parameters

Symbol	Value	Units	Description
dd_{oG}	1.400E-04	m	oil droplet size to be removed from gas
dd_{oW}	5.000E-04	m	oil droplet size to be removed from water
dd_{wO}	9.500E-04	m	water droplet size to be removed from oil
Q_G	0.0456	m³/s	gas flow rate at separator pressure and inlet temperature
Q_O	0.0386	m³/s	oil flow rate at separator pressure and inlet temperature
Q_W	0.0345	m³/s	water flow rate at separator pressure and inlet temperature
μ_G	1.300E-05	Pa.s	gas viscosity at separator pressure and inlet temperature
μ_O	5.250E-03	Pa.s	oil viscosity at separator pressure and inlet temperature
μ_W	4.300E-04	Pa.s	water viscosity at separator pressure and inlet temperature
ρ_G	49.7	kg/m³	gas density at separator pressure and inlet temperature
ρ_O	833.5	kg/m³	oil density at separator pressure and inlet temperature
ρ_W	1030	kg/m³	water density at separator pressure and inlet temperature
σ	0.02	N/m	gas/oil surface tension at separator pressure and inlet temperature
L_{IN}	1.0000	m	location of fluid-saturating the half-bored of the inlet-in line
R_{IN}	0.08	-	height of top of inlet nozzle expressed as proportion of $d_i D_i$

Design Constraints

Symbol	Value	Units	Description
D_i^{MAX}	4	m	maximum practicable internal diameter of vessel
L^{MAX}	20	m	maximum practicable overall lifting of vessel
R_S^{MAX}	7	-	maximum allowable slenderness ratio
Δh_{CON}^{MIN}	0.1	m	minimum height difference for level control between Normal Level / Level A and Level B: both liquid phases
Δt_{CON}^{MIN}	30	s	minimum holdup time between Normal Level / Level A and Level B: both liquid phases
Δh_{HLL}^{MIN}	0	m	minimum height allowance for deaerating between High Level HLL and Level B
Δh_{HHLL}^{MIN}	0.1	m	minimum height difference between HHLL and edge of inlet device, to avoid transient flooding of inlet device following HLL trip
$\Delta t_{HLL,HHLL}^{MIN}$	10	s	minimum holdup time between HLL and top of oil outlet/vortex breaker to avoid transient gas blowby following LLL trip
V_{LIG}^{MIN}	0	m³	minimum slug volume allowance between Normal Level NLL and HLL
V_{SURGE}^{MIN}	0	m³	minimum surge volume allowance between Normal Level NLL and LLL
$t_{RES,O}^{MIN}$	0	s	minimum retention time for oil; from flow straightening baffle to weir; between NLL and NLL
$t_{RES,W}^{MIN}$	0	s	minimum retention time for water; from flow straightening baffle to weir; between NLL and NLL
U_O^{MAX}	0.2	m/s	maximum allowable axial velocity of oil phase, based on cross-sectional area between the weir and NLL and NLL
U_W^{MAX}	0.1	m/s	maximum allowable axial velocity of water phase, based on cross-sectional area between the weir and NLL and NLL

FIGURE 9.93 - PHASE SEPARATOR-INPUT DATA FOR LINE SIZING CALCULATION

Back to 6.3

Results of Separator Design Calculations - Live Field Oil Fluids

D_i (m)	DL_{KOM} (m)	DL_{KMI} (m)	DL_{KOSS} (m)	LL (m)	LL_{S-S} (m)	LL_{eff} (m)	LL_{HHL} (m)	LL_{OL} (m)	V_L (m³)
4.0000	0.6992	0.5390	0.3855	16.0888	14.4888	11.0300	0.0988	1.0080	193.7885

$t_{RES,L}$ (s)	$t_{RES,W}$ (s)	V_{SURGE} (m³)	V_{SURGE} (m³)	$U_{G,i}$ (m/s)	$U_{W,i}$ (m/s)	$\Delta t_{S,M-L}$ (s)	$\Delta h_{S,M-L}$ (m)	$\Delta t_{E,M-W}$ (s)	$\Delta h_{E,M-W}$ (m)
4.0000									
			11.5008	0.0998		8	0.1000		
93		11.5008			0.1029	8		30	0.1886
								30	0.1848
								30	0.1848
						18		30	0.1885
								30	0.1939
					0.1000			30	0.1989
	120							30	0.2008
								72	0.5158

h_T (m)	
4.0000	$h_{T_{V,L}}$
2.5038	$h_{INV_{IN}}$
2.4008	$h_{MHH_{HHL}}$
2.2222	$h_{NH_{HL}}$
2.0307	$h_{NH_{NH}}$
1.8544	$h_{LH_{LL}}$
1.6668	$h_{LH_{LLL}}$
1.6668	$h_{MHoil}/h_{MHwater}$
1.4776	$h_{NH_{HL}}$
1.2727	$h_{NH_{NH}}$
1.0707	$h_{LH_{LH}}$
0.4898	$h_{LH_{LLL}}$

Calculation of Maximum Allowable Axial Gas Velocity @ HHL

U_G^{MAX} (m/s)	U_G^T (m/s)	U_G^{MAX} (m/s)	$N_R N_C$	$Re_P Re_C$	$P_{HL}Re_{HL}$	$D_{HL}Re_{HHL}$
1.0800	0.9939	0.9939	3.28E-02	6.08E-04	7.1E-05	4.45E-50

Trial Values

Trial Values	h_i (m)
$hh_{T_{V,L}}$	4.0000
$h_{MHH_{HHL}}$	2.4008
$h_{NH_{HL}}$	2.2222
$h_{NH_{NH}}$	2.0307
$h_{LH_{LL}}$	1.8544
$h_{LH_{LLL}}$	1.6668
$h_{MHoil}/h_{MHwater}$	1.6668
$h_{NH_{HL}}$	1.4776
$h_{NH_{NH}}$	1.2727
$h_{LH_{LH}}$	1.0707

Calculate the Terminal Velocity of Oil Droplets Settling from Gas

Goal Seek Calculation | **Velocity Calculation**

C_DC_{DOG} Trial	C_DC_{DOG} calc'd	C_DC_{DOG} Trial/Calc'd	Re_PRe_{OG}	F_DF_{DOG} (N)	U_tU_{tOG} (m/s)
0.9090	0.9090	1.0000	95.9797	1.10E-08	0.1079

Calculated U_tU_{tOG} Values (m/s)

@ V_GV_{GHL}	@ V_GV_{HL}
0.7079	0.9484

Calculate the Terminal Velocity of Oil Droplets Rushing from Water

Goal Seek Calculation | **Velocity Calculation**

C_DC_{DOW} Trial	C_DC_{DOW} calc'd	C_DC_{DOW} Trial/Calc'd	Re_PRe_{OW}	F_DF_{DOW} (N)	U_tU_{tOW} (m/s)
1.5656	1.5656	1.0000	34.9505	1.27E-07	0.0028

Calculated U_tU_{tOW} Values (m/s)

@ V_GV_{GHL}	@ V_GV_{HL}
4.92E-02	4.26-26

Calculate the Terminal Velocity of Water Droplets Settling from Oil

Goal Seek Calculation | **Velocity Calculation**

C_DC_{DWO} Trial	C_DC_{DWO} calc'd	C_DC_{DWO} Trial/Calc'd	Re_PRe_{WO}	F_DF_{DWO} (N)	U_tU_{tWO} (m/s)
12.47	12.48	1.000	2.366	9.02E-07	0.0065

Calculated U_tU_{tWO} Values (m/s)

@ V_GV_{GHL}	@ V_GV_{HL}
6647	6638

LL_{EFF}^{MIN}	6647
L_{EFF}^{TOTAL}	11033

FIGURE 20: 3-PHASE SEPARATOR – RESULTS OF DESIGN CALCULATIONS WITH LIVE FIELD OIL FLUIDS

Design Parameters

Symbol	Value	Unit	Description
d_{OG}	1.40E-04	m	oil droplet size to be removed from gas
d_{OW}	5.00E-04	m	oil droplet size to be removed from water
d_{WO}	9.60E-04	m	water droplet size to be removed from oil
Q_G	0.4556	m^3/s	gas flowrate at separator pressure and temperature
Q_O	0.5111	m^3/s	oil flowrate at separator pressure and temperature
Q_W	0.0797	m^3/s	water flowrate at separator pressure and temperature
μ_G	1.30E-05	Pa.s	gas viscosity at separator pressure and temperature
μ_O	5.25E-03	Pa.s	oil viscosity at separator pressure and temperature
μ_W	4.30E-04	Pa.s	water viscosity at separator pressure and temperature
ϱ_G	49.7	kg/m^3	gas density at separator pressure and temperature
ϱ_O	831.5	kg/m^3	oil density at separator pressure and temperature
ϱ_W	1030	kg/m^3	water density at separator pressure and temperature
σ	0.02	N/m	gas/oil surface tension at separator pressure and temperature
L_{IN}	1.000	m	location of flow-straightening baffle, inboard from inlet tan line
R_{IN}	0.8	-	height of top of inlet nozzle expressed as a proportion of D_i

Design Constraints

Symbol	Value	Unit	Description
D_i^{MAX}	4	m	maximum practicable internal diameter of vessel
L^{MAX}	20	m	maximum practicable overall length of vessel
R_S^{MAX}	7	-	maximum allowable slenderness ratio
Δh_{CON}^{MIN}	0.1	m	minimum height difference for level control between Normal Level / Level Alarm / Level Trip, both liquid phases
Δt_{CON}^{MIN}	30	s	minimum holdup time between Normal Level / Level Alarm / Level Trip, both liquid phases
Δh_{FOAM}^{MIN}	0	m	minimum height allowance for foaming, between HLL and HHLL
$\Delta h_{SAF(HHLL)}^{MIN}$	0.1	m	minimum height difference between HHLL and lower edge of inlet device, to avoid transient flooding of inlet device following HHLL trip
$\Delta t_{SAF(LLLL)}^{MIN}$	10	s	minimum holdup time between LLLL and top of oil outlet vortex breaker, to avoid transient gas blowby following LLLL trip
V_{SLUG}^{MIN}	0	m^3	minimum slug volume allowance between NLL and HLL
V_{SURGE}^{MIN}	0	m^3	minimum surge volume allowance between NLL and LLL
$t_{RES,O}^{MIN}$	0	s	minimum retention time for oil, from flow-straightening baffle to weir, between NLL and NIL
$t_{RES,W}^{MIN}$	0	s	minimum retention time for water, from flow-straightening baffle to weir, between NIL and bottom of vessel
U_O^{MAX}	0.2	m/s	maximum allowable axial velocity of oil phase, based on cross-sectional area between NLL and NIL
U_W^{MAX}	0.1	m/s	maximum allowable axial velocity of water phase, based on cross-sectional area between NIL and bottom of vessel

FIGURE 21: 3-PHASE SEPARATOR – INPUT DATA FOR SIZING CHECK WITH EARLY FIELD LIFE FLUIDS

Back to 6.4

Results of Separator Design Calculations - Check on Suitability of Late Life Separator Design for Early Life Fluids

D_i (m)	$D_{N,IN}$ (m)	$D_{N,O}$ (m)	$D_{N,W}$ (m)	$D_{N,GAS}$ (m)	L (m)	L_{S-S} (m)	L_{S-S}/D_i	L_{EFF} (m)	L_W (m)	L_O (m)	V (m³)
4.000	0.692	0.591	0.489	0.305	14.098	12.098	3.024	8.938	0.978	1.182	168.782

$t_{RES,O}$ (s)	$t_{RES,W}$ (s)	V_{SURGE} (m³)	V_{SLUG} (m³)	U_G (m/s)	U_O (m/s)	U_W (m/s)	Δt_{SAF} (s)	Δh_{SAF} (m)	Δh_{FOAM} (m)	Δt_{CON} (s)	Δh_{CON} (m)	h (m)	
												4.000	h_{TV}
							9	0.100				2.508	h_{IN}
				0.098					0.284	30	0.284	2.408	h_{HHLL}
			15.333							30	0.281	2.124	h_{HLL}
59		15.333			0.167					30	0.284	1.844	h_{NLL}
										30	0.295	1.560	h_{LLL}
							10					1.264	h_{LLLL}
										53	0.100	1.242	h_{HHLL}/h_{WEIR}
										52	0.100	1.142	h_{HIL}
	324					0.031				50	0.100	1.042	h_{NIL}
										199	0.453	0.942	h_{LIL}
												0.489	h_{LLIL}

Trial Values

	h (m)
h_{TV}	**4.000**
h_{HHLL}	**2.408**
h_{HLL}	**2.124**
h_{NLL}	**1.844**
h_{LLL}	**1.560**
h_{LLLL}	**1.264**
h_{HHLL}/h_{WEIR}	**1.242**
h_{HIL}	**1.142**
h_{NIL}	**1.042**
h_{LIL}	**0.942**

Calculated Terminal Velocity of Oil Droplets Settling from Gas

Goal Seek Calculation			Velocity Calculation			Calculated $L_{EFF,G}$ Values (m)	
$C_{D,OG}^{Trial}$	$C_{D,OG}^{calc'd}$	$(Trial/Calc'd)$	$Re_{t,OG}$	$F_{d,OG}$ (N)	$U_{t,OG}$ (m/s)	@ h_{LLL}	@ h_{HLL}
0.90	0.90	1.00	95.97	1.10E-08	0.179	0.77	0.82

Calculated Terminal Velocity of Oil Droplets Rising in Water

Goal Seek Calculation			Velocity Calculation			Calculated $L_{EFF,W}$ Values (m)	
$C_{D,OW}^{Trial}$	$C_{D,OW}^{calc'd}$	$(Trial/Calc'd)$	$Re_{t,OW}$	$F_{d,OW}$ (N)	$U_{t,OW}$ (m/s)	@ h_{LIL}	@ h_{HLL}
1.56	1.56	1.00	34.05	1.27E-07	0.028	1.17	1.08

Calculated Terminal Velocity of Water Droplets Settling from Oil

Goal Seek Calculation			Velocity Calculation			Calculated $L_{EFF,O}$ Values (m)	
$C_{D,WO}^{Trial}$	$C_{D,WO}^{calc'd}$	$(Trial/Calc'd)$	$Re_{t,WO}$	$F_{d,WO}$ (N)	$U_{t,WO}$ (m/s)	@ h_{LLL}	@ h_{HLL}
12.47	12.48	1.00	2.36	9.02E-07	0.016	8.94	8.62

Calculation of Maximum Allowable Axial Gas Velocity @ HHLL

U_G^{MAX} (m/s)	ΔU^{MAX} (m/s)	N_μ	Re_f	P_{HHLL}	$D_{H,HHLL}$
1.090	0.993	3.20E-02	6.80E+04	7.105	4.450

L_{EFF}^{MIN}	8.94
L_{EFF}^{TOTAL}	**8.94**

FIGURE 22: 3-PHASE SEPARATOR – RESULTS OF SIZING CHECK WITH EARLY FIELD LIFE FLUIDS

7: NOMENCLATURE

Level Control and Trips

LLIL	low-low interface level (trip)
LIL	low interface level
NIL	normal interface level
HIL	high interface level
HHIL	high-high interface level (trip)
LLLL	low-low liquid (oil) level (trip)
LLL	low liquid (oil) level
NLL	normal liquid (oil) level
HLL	high liquid (oil) level
HHLL	high-high liquid (oil) level (trip)

General Terminology

A	area	m^2
C_1	model/geometry factor in erosion calculation	-
C_D	drag coefficient	-
C_{unit}	unit conversion factor in erosion calculation	-
d	diameter of particle/droplet/bubble	m
D	diameter of nozzle	m
D_H	hydraulic diameter	m
D_i	internal diameter of vessel	m
E	rate of erosion	mm/year
ESD	emergency shut down	-
F	drag force	N

General Terminology, cont.

$F(a)$	sand impact angle factor in erosion calculation	-
g	acceleration due to gravity	m/s²
G	corrections factor in erosion calculation	-
GF	geometry factor in erosion calculation	-
h	liquid height referred to vessel bottom	m
J	velocity head	Pa
K	material coefficient in erosion calculation	m/s
L	length	m
m	mass flow rate	kg/s
NPSH	pump net positive suction head	m
N_μ	viscosity number	-
P	wetted perimeter	m
ppmM	parts per million, mass basis	-
Q	volumetric flow rate	m³/s
R	radius of curvature of pipe divided by NPS	-
Re	Reynolds number	-
Re_f	Reynolds film number	-
R_{IN}	proximity of inlet nozzle to vessel head	-
R_S	slenderness ratio	-
SIL	safety integrity level	-
t	time	s
T	temperature	°C

Subscripts & Superscripts

CON	control
d	droplet/bubble
E	erosional
EFF	effective
G	gas
GAS	gas outlet nozzle
go	gas bubbles, oil phase continuous
H	vessel head
IN	inlet
L	liquid
M	mixed (phases)
MAX	maximum allowable
MIN	minimum allowable
N	nozzle
O	oil
og	oil droplets, gas phase continuous
ow	oil droplets, water phase continuous
P	pipe
RES	residence (time)
S	sand
SAF	safety
S-S	seam to seam

Subscripts & Superscripts, cont.

TV	top of vessel	
V	velocity in nozzle	
VH	velocity head in nozzle	
W	water	
wo	water droplets, oil phase continuous	

Greek Symbols

Δ	differential	-
μ	viscosity	Pa.s
ρ	density	kg/m3
σ	gas-liquid interfacial tension	N/m

Other Symbols

$|x - y|$ magnitude of $x - y$

8. REFERENCES

American Petroleum Institute (2008): *Specification for Oil and Gas Separators - API Specification 12J* (8th Ed.). Washington, DC: API Publishing Services

American Petroleum Institute (1991): *Recommended Practice for Design and Installation of Offshore Production Platform Piping Systems - API Recommended Practice 14E (RP 14E)* (5th Ed.). Washington, DC: API Publishing Services

Arnold, K. and Koszela, P. (1990): "Droplet-Settling vs. Retention-Time Theories for Sizing Oil/Water Separator", SPE Production Engineering, February 1990

Arnold, K. and Stewart, M. (2008): *Surface Production Operations Design of Oil Handling Systems and Facilities* (3rd Ed.). Burlington, MA: Elsevier Inc.

Arntzen, R.: "Level Design and Control in Gravity Separators", Oil and Gas Facilities, September 2016

DNV GL AS: *Recommended Practice, Managing Sand Production and Erosion - DNVGL-RP-O501* (Edition August 2015). Oslo

GPSA (2012): *"Engineering Data Book"*, (13th Ed.). Tulsa, OK: Gas Processors Suppliers Association

Grødal, E.O. and Realff, M.J.: "Optimal Design of Two- and Three-Phase Separators: A Mathematical Programming Formulation", Paper SPE 56645, presented at the 1999 SPE Annual Technical Conference and Exhibition, Houston, TX, USA, Oct. 3-6, 1999

Hansen, E.W.M.; Heitmann, H.; Laska, B., and Loes, M.: "Numerical Simulation of Fluid Flow Behavior Inside, and Redesign of a Field Separator", Proc., 6th International Conference on Multiphase Production, Cannes, France, 19-21 June 1993, 117-129.

Hoffmann, A.E.; Crump, J.S. and Hocott, C.R.: "Equilibrium Constants for a Gas-Condensate System", Petroleum Transactions, AIME, Vol. 198, 1953

Ishii, M. and Grolmes, M.A.: "Inception Criteria for Droplet Entrainment in Two-Phase Concurrent Film Flow", AIChE J., March 1975

Laleh, A.P.; Svrcek, W.Y. and Monnery, W.D.: "Computational Fluid Dynamics-Based Study of an Oilfield Separator - Part II: An Optimum Design", Oil and Gas Facilities, February 2013

Monnery, W.D. and Svrcek, W.Y.: "Successfully Specify Three-Phase Separators," Chemical Engineering Progress, September 1994

NORSOK (2014): *Process System Design* – NORSOK Standard P-002 (1st Ed.): 1326 Lysaker, Norway: Standards Norway

Rochelle, S.G. and Briscoe, M.T.: "Predict and Prevent Air Entrainment in Draining Tanks", Chemical Engineering, November 2010

Viles, J.C.: "Predicting Liquid Re-Entrainment in Horizontal Separators," Journal of Petroleum Technology, May 1993

9. ABOUT THE AUTHOR

In 1980 the author graduated BSc (Hons) in Chemical and Process Engineering from Heriot-Watt University, Edinburgh, UK.

During the next 35 years he worked as a Process engineer in the Oil and Gas sector, on design projects and in operations support.

www.ingramcontent.com/pod-product-compliance
Lightning Source LLC
Chambersburg PA
CBHW051916210526
45473CB00006B/2034